ATMOSPHERIC DEGRADATION and CORROSION CONTROL

CORROSION TECHNOLOGY

Editor
Philip A. Schweitzer, P.E.
Consultant
Fallston, Maryland

1. *Corrosion and Corrosion Protection Handbook: Second Edition, Revised and Expanded*, edited by Philip A. Schweitzer
2. *Corrosion Resistant Coatings Technology*, Ichiro Suzuki
3. *Corrosion Resistance of Elastomers*, Philip A. Schweitzer
4. *Corrosion Resistance Tables: Metals, Nonmetals, Coatings, Mortars, Plastics, Elastomers and Linings, and Fabrics: Third Edition, Revised and Expanded (Parts A and B)*, Philip A. Schweitzer
5. *Corrosion-Resistant Piping Systems*, Philip A. Schweitzer
6. *Corrosion Resistance of Zinc and Zinc Alloys*, Frank C. Porter
7. *Corrosion of Ceramics*, Ronald A. McCauley
8. *Corrosion Mechanisms in Theory and Practice*, edited by P. Marcus and J. Oudar
9. *Corrosion Resistance of Stainless Steels*, C. P. Dillon
10. *Corrosion Resistance Tables: Metals, Nonmetals, Coatings, Mortars, Plastics, Elastomers and Linings, and Fabrics: Fourth Edition, Revised and Expanded (Parts A, B, and C)*, Philip A. Schweitzer
11. *Corrosion Engineering Handbook*, edited by Philip A. Schweitzer
12. *Atmospheric Degradation and Corrosion Control*, Philip A. Schweitzer

ADDITIONAL VOLUMES IN PREPARATION

Environmental Degradation of Metals, U. K. Chatterjee, S. K. Bose, and S. K. Roy

ATMOSPHERIC DEGRADATION and CORROSION CONTROL

Philip A. Schweitzer, P.E.
Consultant
Fallston, Maryland

MARCEL DEKKER, INC. NEW YORK • BASEL

ISBN: 0-8247-7709-3

This book is printed on acid-free paper.

Headquarters
Marcel Dekker, Inc.
270 Madison Avenue, New York, NY 10016
tel: 212-696-9000; fax: 212-685-4540

Eastern Hemisphere Distribution
Marcel Dekker AG
Hutgasse 4, Postfach 812, CH-4001 Basel, Switzerland
tel: 41-61-261-8482; fax: 41-61-261-8896

World Wide Web
http://www.dekker.com

The publisher offers discounts on this book when ordered in bulk quantities. For more information, write to Special Sales/Professional Marketing at the headquarters address above.

Current printing (last digit):
10 9 8 7 6 5 4 3 2 1

PRINTED IN THE UNITED STATES OF AMERICA

Preface

Atmospheric corrosion is a serious worldwide problem affecting people in all walks of life. It is of concern to homeowners, architects, engineers, designers, maintenance personnel, and accountants. All outdoor and indoor materials exposed to the elements are potentially subject to degradation resulting from atmospheric conditions.

Damage caused by atmospheric corrosion can be dangerous as well as costly. Bridges failing as the result of rusting and consequent failure of steel parts; portions of concrete from buildings have spalled and fallen to the ground, endangering passers-by; and essential automobile parts failing from corrosion are all examples of dangers posed by atmospheric degradation. Costly repairs include resurfacing of roadways after corrosion by de-icing compounds, repair and painting of bridges and towers, replacing of fencing, and maintenance of building exteriors, to name a few.

However, steps can be taken to keep corrosion within tolerable limits or to eliminate it completely. This book will provide the reader with an understanding of the factors in and mechanisms of atmospheric corrosion. Also covered are the effects of the atmospheric environment on all the major materials of construction, including various metals, masonry materials, wood, plastics, elastomers, coatings (metallic and organic), and ceramics.

Based on this information, design principles are provided to alleviate the problems of atmospheric degradation and programs are proposed for monitoring the effectiveness of coatings.

It is hoped that the reader will find this book helpful and beneficial.

Philip A. Schweitzer

Contents

1
Introduction

Several years ago the National Institute of Standards and Technology (formerly the National Bureau of Standards) estimated that the cost of corrosion in the United States was in the range of $9 billion to $90 billion. These figures were confirmed by various technical organizations such as the National Association of Corrosion Engineers (NACE) International and others.

Included in this estimate was corrosion attributed to chemical processes; to corrosion of highways and bridges from deicing chemicals; to atmospheric corrosion of steel fences; to atmospheric corrosion of various outdoor structures such as buildings, bridges, towers, automobiles, and ships; and innumerable other applications exposed to the atmospheric environment. It has been further estimated that the cost of protection against atmospheric corrosion is approximately 50% of the total cost of all corrosion protection methods.

Atmospheric corrosion, sometimes referred to as weathering, is a complicated electrochemical process that takes place in corrosion cells consisting of base metal, metallic corrosion products, surface electrolytes, and the atmosphere. The corrosion characteristics of the atmosphere are influenced by many variables including relative humidity, temperature, sulfur dioxide content, hydrogen sulfide content, chloride content, amount of rainfall, dew formation, dust, and even the position of the exposed metal. Geographic location also plays an important part. For example, carbon steel parts left in the desert remain bright and tarnish-free over long periods of time due to the absence of moisture. Also in climates where the air temperature is below the freezing point of water or of aqueous condensation on the metal surface, rusting is negligible because ice is a poor conductor and does not function as an electrolyte.

Atmospheric corrosion depends not only on the moisture content present but on the dust content and the presence of other impurities in the air, all of which have an effect on the condensation of moisture on the metal surface and resulting corrosiveness. The temperature of the air can also be a factor.

All types of corrosion may take place, depending on the specific contaminants present and the materials of construction. General corrosion is the predominant form encountered because of the large quantities of steel used. However, localized forms such as pitting, intergranular attack, and stress corrosion cracking may be encountered with susceptible alloys. Because the electrolyte available consists only of a thin film of condensed or absorbed moisture the possibility of galvanic corrosion is somewhat minimized. However, this cannot be relied on and galvanic corrosion must always be considered in design for atmospheric exposures.

Synthetic materials as well as metals are also subject to atmospheric corrosion depending on the specific synthetic material and the conditions of exposure. Synthetic materials, plastics, and elastomers can be subject to degradation as a result of the action of ozone, oxygen, and sunlight. These three weathering agents can greatly affect the properties and appearance of a large number of synthetic materials. Surface cracking, discoloration of colored stock, and serious loss of tensile strength are the result of this attack. Elastomeric materials may also suffer loss of elongation and other rubber-like properties.

Considering the extent of atmospheric corrosion, its cost to society, and the potential danger of failures of bridges and other structures, it is no wonder that much time and effort has been expended to study the causes and effects. These studies have been conducted worldwide, for the most part in the form of field tests at sites in different types of climate. Laboratory investigations have also been conducted using advanced research tools for investigations of atmospheric corrosion and its mechanisms.

I. ATMOSPHERIC TYPES

Since atmospheric corrosion rates are affected by local conditions, atmospheres are classified according to exposure conditions. The major categories, based on potential corrosion rates, are:

Rural
Urban
Industrial

Marine
Indoor

The first four can be further subdivided into arctic, temperate, and tropical (wet or dry) since the difference in temperature, humidity, and seawater composition all have a bearing on the corrosion rates. However, we are primarily concerned with conditions in the temperate category.

A rural atmosphere is generally that of an inland farm with little or no heavy manufacturing operations. In such areas the problems of atmospheric corrosion will be somewhat diminished. The same will apply in very dry climatic conditions.

Urban atmospheres, even when free from industrial pollution, are subject to pollution from road traffic and the burning of fossil fuels. The former produces oxides of nitrogen, which may be oxidized to nitric acid. The latter has the potential of generating sulfur dioxide, which is converted to sulfuric and sulfurous acid in the presence of moisture.

In an industrial atmosphere all types of contamination by sulfur in the form of sulfur dioxide or hydrogen sulfide are important. The burning of fossil fuels generates a large amount of sulfur dioxide, which is converted to sulfuric and sulfurous acids in the presence of moisture. Combustion of these fossil fuels and hazardous waste products should produce only carbon dioxide, water vapor, and inert gas as combustion products. This is seldom the case. Depending on the impurities contained in the fossil fuel, the chemical composition of the hazardous waste materials incinerated, and the combustion conditions encountered, a multitude of other compounds may be formed.

In addition to the most common contaminants previously mentioned, pollutants such as hydrogen chloride, chlorine, hydrogen fluoride, and hydrogen bromide are produced as combustion products from the burning of chemical waste. When organophosphorous compounds are incinerated, corrosive phosphorous compounds are produced. Chlorides are also a product of municipal waste incinerators.

Energy production leads to the formation of oxides of nitrogen, which may be oxidized to nitric acid. This reaction has a very low rate; therefore, in the vicinity of the emission source the concentration of nitric acid and nitrates is very low. The greatest effect will occur at some distance from the emission source.

The antipollution regulations that have been enacted do not prevent the escape into the atmosphere of quantities of these materials sufficient to prevent corrosion problems. The corrosivity of an industrial atmosphere diminishes with increasing distance from the area.

Marine environments are subject to chloride attack resulting from the deposition of fine droplets of crystals formed by evaporation of spray that has been carried by the wind from the sea. The quantity of chloride deposition from a marine environment is directly proportional to the distance from the shore. The closer to the shore, the greater the deposition and corrosive effect. The atmospheric test station at Kure Beach, North Carolina shows that steels exposed 80 feet from the ocean corrode 10–15 times faster than steels exposed 800 ft from the ocean. The distance from the shore at which marine atmospheres cease depends on the prevailing winds, the distance they travel over at speed, and the height of the local surf. Mixtures of marine and industrial effects are common in many areas with each type of contaminant aggravating the other.

In addition to these general air contaminants there may also be specific pollutants found in a localized area. These may be emitted from a manufacturing operation on a continuous or spasmodic basis and can result in a much more serious corrosion problem than that caused by the presence of general atmospheric pollutants.

Because of these varying conditions a material that is resistant to atmospheric corrosion in one area might not be satisfactory in another area. For example, galvanized iron is perfectly suitable for application in a rural atmosphere but is not suitable when exposed to industrial atmospheres.

To compound the problem there is no clear line of demarcation between these atmospheric types. In many cases there is no "pure" rural or urban area. Contamination from industrial or marine areas can find its way into these areas based on the prevailing winds and other atmospheric conditions.

Indoor atmospheres might be free of corrosion in "clean rooms" or subject to severe corrosion as around a prickling bath in a steel mill.

Atmospheric conditions should be defined in terms of temperature, humidity, and contaminants, as well as their corrosivity to the specific materials of construction being considered. In addition to the general atmospheric conditions, special conditions such as cooling tower drift or spray, spills, or releases of water or chemicals should not be overlooked and must be taken into account.

2
Factors Affecting Atmospheric Corrosion

Atmospheric corrosion is an electrochemical process and as such depends on the presence of an electrolyte. The usual electrolyte associated with atmospheric corrosion is water resulting from rain, fog, dew, melting snow, or high humidity. Since an electrolyte is not always present, atmospheric corrosion is considered a discontinuous process. Corrosion takes place only during the time of wetness. It can be described by the equation:

$$K = \sum_{l}^{n} t_n v_k(n)$$

where

K = accumulated corrosion effect
t_n = time of wetness (the period with an electrolyte layer on the surface)
v_k = average corrosion rate during the individual periods of wetness

The total time of wetness, the composition of the electrolyte, and the temperature determines the corrosion rate. Factors affecting the time of wetness and the composition of the electrolyte film will be discussed later.

In addition to the above there are other factors that affect the corrosion rate. Included are initial exposure conditions, sample mass, orientation, extent of sheltering, wind velocity, nature of the corrosion products formed, and pollutants present (both known and unknown).

I. TIME OF WETNESS

The term "time of wetness" refers to the length of time during which the metal surface is covered by a film of water that renders significant atmospheric corrosion possible. The actual time of wetness will vary with climatic conditions at the location. It is dependent on the relative humidity of the atmosphere (being greatest when the relative humidity is ≥80%); the temperature of the air and the metal surface above 32°F/0°C; the duration and frequency of rain, fog, dew, and melting snow; as well as the hours of sunshine and the wind speed.

Moisture on the surface of the metal resulting from high humidity of the atmosphere and the chemical and physical properties of the corrosion products produces an adsorption layer of water. The metal surface may be wetted if hygroscopic salts deposited or formed by corrosion absorb water from the atmosphere (see Chapter 3). Phase layers of water are the result of rain, fog, wet or melting snow, or dew formed by condensation on cold metallic surfaces.

The total time of wetness (T_{tw}) may be divided into the periods when the metal is moistened due to adsorption of water vapor on the surface (T_{ad}) and the periods when the surface is covered by a phase layer of water (T_{ph}) resulting from rain, fog, dew, or wet or melting snow:

$$T_{tw} = T_{ad} + T_{ph}$$

It is very difficult to distinguish experimentally between the two categories of time of wetness because there is no sharp boundary.

A. Measurement of Time of Wetness

The time of wetness can be calculated either by meteorological measurements of temperature and relative humidity or by electrochemical cells. For practical purposes, meteorological measurements are used when the relative humidity is ≥80% at temperatures >32°F/0°C to determine the time of wetness. However, the time of wetness determined in this manner may not be the same as the actual time of wetness because wetness is also influenced by the type of metal, pollution of the atmosphere, presence of corrosion products, and the degree of coverage against rain. Even so, the results from these measurements usually show a good correlation with corrosion data from field tests under ordinary outdoor conditions.

Electrochemical cells may also be used to directly measure the time of wetness. These cells consist of thin, electrically separated, metal electrodes (0.5–1 mm). Thin insulators separating the electrodes are 0.1–0.2 mm in size. When the aqueous layer is thick enough to accelerate atmospheric corrosion, a current or electromotive force is generated. The time of wetness is defined as the time when the cell current or the electromotive force exceeds a certain value. Measurement of the time of wetness from the cell will not necessarily be unambiguous because its result is dependent on such factors as cell design, criteria for determining the time of wetness based on current or electromotive responses, cell dimensions, and the presence of corrosion products. Most of the electrochemical techniques indicate primarily the time of wetness caused by phase layers of electrolyte. These values are usually lower than the values from calculations using meteorological data. In the temperate zones values of 1000–2700 h/year are often reported.

II. RAIN

Precipitation in the form of rain has a dual effect on the atmospheric corrosion of metals. On the one hand, it affects corrosion by forming a phase layer of moisture on the material surface and by adding corrosion stimulators in the form of H^+ and SO_4^{2-}. On the other hand, it tends to wash away pollutants deposited on the surface during the preceding dry period. Whereas the first two processes promote corrosion, the third tends to decrease corrosion. The significance of the two latter processes is dependent on the ratio of dry and wet deposition of pollutants.

The corrosion on the skyward side of metal plates in a strongly polluted atmosphere will be substantially lower than that on the downward side. In a strongly polluted atmosphere, where dry deposition is considerably greater than wet deposition of sulfur pollutants, the washing effect of rain will predominate. In areas that are not heavily polluted the situation will be reversed. The corrosive action of the rain is more important; consequently, the skyward side of metal plates will have a higher corrosion rate.

The pH value of the rain dictates the degree of corrosion that will take place. When the pH value of the rainfall approaches 4 or less, accelerated corrosion resulting from the dissolution of the protective layers of basic carbonates or sulfates as on zinc and copper will take place.

III. DEW

Dew is an important source of atmospheric corrosion—more so than rain—and particularly under sheltered conditions. Dew forms when the temperature of the metal falls below the dew point of the atmosphere. This can occur outdoors during the night when the surface temperature of the metal is lowered as a result of radiant heat transfer between the metal and the sky. It is also common for dew to form during the early morning hours when the air temperature rises more quickly than the metal temperature. Dew may also form when metal products are brought into a warm storage area after cold shipment.

Under sheltered conditions dew is an important cause of corrosion. The high corrosivity of dew is the result of several factors:

1. Relatively speaking, the concentration of contaminants in dew is higher than in rainwater, which leads to lower pH values. Heavily industrialized areas have reported pH values of dew in the range of 3 or lower.
2. The washing effect, which occurs with rain, is usually slight or negligible. With little or no run-off, the pollutants remain in the electrolyte and continue their corrosive action. As the dew dries these contaminants remain on the surface to repeat their corrosive activity with subsequent dew formation.

The amount of water on a surface covered by dew is approximately 10 g/m^2, which is considerably more than on surfaces covered by adsorption layers. In heavily polluted areas dew pH values of 3 or lower have been found on inert substrates such as painted roofs, where dry deposition between periods of rain have led to the accumulation of corrosive, fairly acid pollutants. Concentrations of 0.35 g/L of chlorides and 0.20 g/L of sulfates have been reported. These values are approximately 100 times higher than those formed in rainwater at the same location. This extreme corrosivity formed by dew condensation may be seen in the effect of run-off of condensed water from painted roofs. Such run-off may cause perforation of gutters of galvanized steel plate within 3 years.

IV. FOG

In areas of high pollution, fog droplets will have a high acidity and contain high concentrations of sulfates and nitrates. In areas of California the pH of the fog water has been found to be in the range of 2.2–4.0.

V. DUST

On a weight basis, in many locations dust is the primary air contaminant. When in contact with metallic surfaces and combined with moisture, dust can promote corrosion by forming galvanic or differential aeration cells that, because of their hygroscopic nature, form an electrolyte on the surface. Suspended particles of carbon and carbon compounds, metal oxides, sulfuric acid, ammonium sulfate, sodium chloride, and other salts will be found in industrial atmospheres. It is these materials, when combined with moisture, that initiate corrosion.

The settled dust may promote corrosion by absorbing sulfur dioxide and water vapor from the air. Hygroscopic salts such as chlorides or sulfates form a corrosive electrolyte on the surface. Carbonaceous particles can start the corrosion process by forming cathodes in microcells with a steel surface.

Dust-free air therefore is less likely to cause corrosion.

VI. TEMPERATURE

The overall effect of temperature on corrosion rates is complex. During long-term exposure in a temperate climatic zone, temperature appears to have little or no effect on the corrosion rate. As the temperature increases the rate of corrosive attack increases as a result of an increase in the rate of electrochemical and chemical reactions as well as the diffusion rate. Consequently, under constant humidity conditions, a temperature increase will promote corrosion. Conversely, an increase in temperature can cause a decrease in the corrosion rate by causing a more rapid evaporation of the surface moisture film created by rain or dew. This reduces the time of wetness, which in turn reduces the corrosion rate. In addition, as the temperature increases, the solubility of oxygen and other corrosive gases in the electrolyte film is reduced.

When the air temperature falls below 32°F/0°C the electrolyte film might freeze. As freezing occurs there is a pronounced decrease in the corrosion rate, which is illustrated by the low corrosion rates in the subarctic and arctic regions.

In general, temperature is a factor influencing corrosion rates, but it is of little importance except under extreme conditions.

VII. INITIAL EXPOSURE CONDITIONS

Atmospheric corrosion rates proceed through three stages or periods: the induction period, the transition period, and the stationary period. During the induction period the metal is covered with a spontaneously formed oxide and the aqueous layer. This oxide provides some degree of protection, depending on the metal and the aggressiveness of the atmosphere. During the transition period the oxide layer transforms into a fully developed layer of corrosion products. The final or stationary period is characterized by the surface being fully covered by corrosion products and eventually reaching constant properties with respect to chemical composition and stationary corrosion rates. The more aggressive the exposure conditions the shorter will be the two initial periods. For example, in a benign (indoor) atmosphere these initial periods for steel will last for years, but in highly polluted industrial areas these initial periods might last only a few months.

The initial exposure conditions have a very marked influence on the subsequent corrosion rate. During the first days of exposure, wet conditions (caused by high relative humidity or rainfall) cause higher corrosion rates than dry conditions. These effects will vary from one material to another. For example, zinc is more sensitive than steel. Differences are explained by the fact that different materials form different corrosion products with different protective properties. A wide variety of structurally related corrosion products can be formed on zinc, the nature of which depends on initial exposure conditions. The seasonal dependence on the concentrations of peroxide and ozone in the atmosphere might also be a contributing factor.

During the third period of exposure constant corrosion rate is established; thus the terminology "stationary period."

VIII. SHELTERING

The design of the system should eliminate all possible areas where water, dirt, and corrosion products can accumulate. The degree of shelter from particle fallout and rainfall can affect initial and ultimate corrosion rates. When pockets are present the time of wetness increases, which leads to corrosive conditions. The design should make maximum use of exposure to the weather. Sheltering from rain should be avoided.

IX. WIND VELOCITY

Wind speed and the type of wind flow have a pronounced effect on the atmospheric corrosion rate. This is illustrated by the dry deposition velocity, which is defined as the ratio of deposition rate of any gaseous compound and the concentration of that compound in the atmosphere. The factors controlling dry deposition are composed of aerodynamic and surface processes. Aerodynamic processes are related to the actual depletion of the gaseous constituent (e.g., SO_2) in the atmospheric region adjacent to the aqueous phase and the ability of the system to add new SO_2 to the region. This ability depends on the actual wind speed, type of wind flow, and the shape of the sample.

Surface processes are dependent on the ability of the aqueous phase to accept SO_2. Greater amounts of SO_2 can be accepted as the thickness of the aqueous phase increases.

The dry deposition velocity (V_d) can be expressed as the inverse of the sum of two resistances: the aerodynamic resistance (R_a) and the surface resistance (R_s).

$$V_d = \frac{1}{R_a + R_s}$$

Under most atmospheric conditions the dry deposition velocity is the combined effect of both resistances. In atmospheric exposures having highly turbulent and well-mixed air flow, $R_a \approx 0$, and dry deposition velocity is solely dependent on the surface processes.

X. NATURE OF CORROSION PRODUCTS

The atmospheric corrosion rate is influenced by the nature of the specific corrosion products formed, some of which have a greater protective ability than others. The composition of the corrosion products formed is dependent on the specific dissolved metal ions and the access to anions dissolved in the aqueous layer.

Depending on the rate of formation and the rate of crystallization, the product may be amorphous or crystalline. Aging or slow growth of amorphous phases may result in a transition from the amorphous to the crystalline state. This process can occur through slow transformation in the solid state or through dissolution-precipitation processes. This is illustrated in the transition from amorphous to crystalline basic nickel sulfate, the former being less corrosion-resistant than the latter.

XI. POLLUTANTS PRESENT

One of the most important factors affecting atmospheric corrosion is the presence of specific pollutants. In areas having low atmospheric pollution, corrosion rates are correspondingly low. The presence of atmospheric pollutants such as the various oxides of nitrogen, sulfur-containing compounds, chlorine-containing compounds, and other less common pollutants will stimulate corrosion.

It has been proven that the following gaseous constituents are of significant importance in contributing to atmospheric corrosion: O_2, H_2O_2, SO_2, H_2S, COS, NO_2, HNO_3, NH_3, HCl, Cl_2, HCHO, and HCOOH. Typical ranges of these materials as found under outdoor and indoor conditions are given in Table 2.1. They may be present as the result of either natural or anthropogenic processes, and may undergo a variety of chemical changes during transport in the atmosphere.

All of the species are reactive and as such have a certain average lifetime that is limited by the ability to react with atmospheric oxidizers, primarily the hydroxyl radical, OH^-, and O_3. OH^- is generated by photoinduced dissociation of O_3 (ozone) and the subsequent reaction of the electronically excited, energy-rich oxygen atom O ('D) and water vapor.

$$O_3 + hv \rightarrow O('D) + O_2 \qquad (\lambda < 310 \text{ nm})$$

$$O('D) + H_2O \rightarrow 2OH\cdot$$

It is possible for the OH· molecules to oxidize several of the species such as SO_2, H_2S, and NO. However, a large portion of the OH· molecules are consumed through reactions with hydrocarbon molecules producing an end-product of $HO_2\cdot$ (the hydroperoxyl radical). This radical converts to hydrogen peroxide (H_2O_2) and O_2 according to:

Table 2.1 Concentration of Selected Gaseous Air Constituents (ppb)

Gas	Outdoor	Indoor
O_3	4–42	3–30
H_2O_2	10–30	5
SO_2	1–65	0.3–14
H_2S	0.7–24	0.1–0.7
NO_2	9–78	1–29
HNO_3	1–10	3
NH_3	7–16	13–259
HCl	0.18–3	0.05–0.18
Cl_2	<0.005–0.08[a]	0.001–0.005
HCHO	4–15	10
HCOOH	4–20	20

[a]Corresponding to 5 wt% HCl.

$$HO_2\cdot + HO_2\cdot \rightarrow H_2O_2 + O_2$$

H_2O_2 is highly soluble in aqueous environments and is a powerful oxidizing agent.

However, the degree to which these pollutants will affect the corrosion rate will be influenced by the other preceding factors that also have an influence on the corrosion rate. The corrosive effect of these pollutants will be dependent on the specific material with which they come into contact. Succeeding chapters deal with the mechanisms of corrosion of these pollutants and their effect on specific materials.

3
Mechanisms of Atmospheric Corrosion of Metals

Atmospheric corrosion takes place by means of an electrochemical process occurring in corrosion cells. A corrosion cell must have the following essential components:

1. A metal anode.
2. A metal cathode.
3. A metallic conductor between the anode and cathode.
4. An electrolyte (water containing conductive salts) in contact with the anode and cathode but not necessarily of the same composition at the two locations.

In addition, oxygen will usually be present as a depolarizing agent. For the cell to function there must be a difference in potential between the anode and cathode. This difference in potential is mainly due to contact between dissimilar metallic conductors or differences in concentration of the solution, generally related to dissolved oxygen in natural waters. Almost any lack of homogeneity on the metal surface or in its environment may initiate attack by causing a differential in potentials that results in more or less localization of corrosion.

Atmospheric corrosion differs from the action that occurs in water or underground in that a plentiful supply of oxygen is always present. In this case, the formation of insoluble films, and the presence of moisture and deposits from the atmosphere become the controlling factors. The presence of contaminants such as sulfur compounds and salt particles also affects the corrosion rate. Nevertheless, atmospheric corrosion is mainly electrochemical, rather than a direct chemical attack by the elements. The anodic and cathodic areas are usually quite small and close together, so that corrosion is apparently uniform rather than in the form of severe pitting as in soil or water. Anodes and cathodes exist on all steel surfaces. Surface imperfections, grain orientation, lack of homogeneity of the metal, variation in the environment, localized shear and torque during manufacture, mill scale, and existing red iron rust will produce anodes and cathodes. It is a relatively complicated system consisting of a metal, corrosion products, surface electrolyte, and the atmosphere. For an electrochemical reaction to proceed two or more electrode reactions must take place: the oxidation of a metal (anodic partial reaction) and the reduction of an oxidizing agent (cathodic partial reaction). In order for the electrode reactions to take place it is necessary that an electrolyte be present.

Most metals, if exposed to the atmosphere at room temperature with virtually no humidity present, will form an oxide film. If the oxide film is stable, the growth rate stops and the oxide film reaches a maximum thickness of 1–5 mm, protecting the metal.

Atmospheric corrosion can fall into two categories: "damp" atmospheric corrosion, which takes place in the presence of water vapor and traces of pollutants, and "wet" atmospheric corrosion, which occurs when rain or other forms of bulk water are present, with pollutants. For atmospheric corrosion to proceed water of some form must be present.

Damp atmospheric corrosion occurs when water is present on the surface of the metal as an aqueous phase layer as caused in humidity, dew, or fog. Wet atmospheric corrosion occurs when bulk water is present, such as rain.

I. DAMP ATMOSPHERIC CORROSION (ADSORPTION LAYERS)

Aqueous phase layers consist of water adsorbed on the metal surface. The amount formed is dependent on the relative humidity of the atmosphere and the chemical and physical properties of the corrosion products.

Water may be absorbed from the atmosphere and wet the metal surface if hygroscopic salts are deposited or formed by corrosion. This absorption will take place when the relative humidity exceeds the critical relative humidity. The value of the critical relative humidity is dependent on the specific metal and the specific metal contaminants. When the relative humidity exceeds the value at which the salt starts to absorb water and dissolve, the corrosion rate increases sharply. This critical relative humidity corresponds to the vapor pressure above a saturated solution of the salt present. Adsorption layers of electrolyte on the surface of the metal may also be the result of capillary condensation.

The corrosion rate is directly influenced by the amount of water present on the corroding surface. Laboratory studies have shown that the corrosion rate above the critical value sharply increases with increasing relative humidity. The amount of water present on a metal surface has been roughly estimated as follows:

Conditions	Amount of water (g/m^2)
Critical relative humidity	0.01
100% relative humidity	1
Covered by dew	10
Wet from rain	100

Atmospheric corrosion resulting from the reaction of water vapor with a metal surface is a serious problem. Most clean metal surfaces will permit the bonding of water in molecular form. The oxygen atom bonds to the metal surface and acts as a Lewis base (donating an electron pair), since the bonding is connected with a net charge transfer from the water molecule to the surface. The water adsorbs on electron-deficient adsorption sites.

It is also possible for water to bond in dissociated form. In this case the driving force is the formation of metal–oxygen or metal–hydroxyl bonds. The end-products formed as a result of the water adsorption are adsorbed hydroxyl, atomic oxygen, and atomic hydrogen. When metal oxides are present, water may adsorb in either dissociative or molecular form. Lattice defect sites seem to facilitate dissociation, as observed, for instance, on monocrystalline TiO_2, NiO, and α-Fe_2O_3. The dissociation of water forms a monomolecular thick film of surface hydroxyl groups that is relatively protective and reduces the subsequent reaction rate of water. The first monolayer of water adsorbed to the hydroxylated oxide surface is highly immobile, whereas the second and third layers are more randomly oriented and less immobile.

Many different metals adsorb water in similar manners, forming metal oxyhydroxides. The exact nature of the oxyhydroxide formed seems to have only a minor influence on the water adsorption phenomena. The quantity of reversibly adsorbed water increases with relative humidity and time. (1) Refer to Table 3.1 for the approximate number of monolayers of water at 77°F/(25°C) and steady-state conditions, which have been experimentally

Table 3.1 Approximate Number of Water Monolayers on Different Metals vs. Relative Humidity

Relative humidity (%)	Number of monolayers
20	1
40	1.5–2
60	2–5
80	5–10

Source: Ref. 1.

determined by the quartz crystal microbalance method on a number of metals.

Gaseous constituents of the atmosphere dissolve in the aqueous layers formed. Corrosive attack is generally found in areas where water adsorption is favored, permitting easy dissolution of the gaseous molecules such as SO_2 and NO_2. The properties of wet atmospheric corrosion are approached when the aqueous films are greater than approximately three monolayers. At this point the relative humidity is close to the "critical relative humidity." At values above the critical relative humidity, atmospheric corrosion rates increase appreciably, whereas below this value atmospheric corrosion is negligible. The critical relative humidity varies for different metals and pollutants.

II. WET ATMOSPHERIC CORROSION (PHASE LAYERS)

Wet atmospheric corrosion results from repeated wet and dry cycles, the presence of pollutants, and the formation of an aqueous layer in which the atmospheric pollutants dissolve. The wet cycles result from dew, fog, rain, or snow. In many cases the dew, fog, rain, or snow may already contain the dissolved corrodent, which then deposits on the surface.

A. Dew

Dew is an important source of atmospheric corrosion—more so than rain—and particularly under sheltered conditions. Dew forms when the temperature of the metal surface falls below the dew point of the atmosphere. This can occur outdoors during the night when the surface temperature of the metal is lowered as a result of radiant heat transfer between the metal and the sky. It is also common for dew to form during the early morning hours when the air temperature rises more quickly than the metal temperature. Dew may also form when metal products are brought into warm storage after cold shipment.

Under sheltered conditions dew is an important cause of corrosion. The high corrosivity of dew is a result of several factors:

1. Relatively speaking, the concentration of contaminants in dew is higher than in rainwater, which leads to lower pH values. Heavily industrialized areas have reported pH values of dew in the range of 3 or lower.
2. The washing effect, which occurs with rain, is usually slight or negligible with dew.

With little or no run-off, the pollutants remain in the electrolyte and continue their corrosive action. As the dew dries, these contaminants remain on the surface to repeat their corrosive activity with subsequent dew formation.

B. Rain

Depending on the conditions, rain can either increase or decrease the effects of atmospheric corrosion. Corrosive action is caused by rain when a phase layer of moisture is formed on the metal surface. Rain creates thicker layers of electrolyte on the surface than dew. The corrosive activity is increased when the rain washes corrosive promoters such as H^+ and SO_4^{2-} from the air (acid rain). Rain has the ability to decrease corrosive action on the surface of the metal as a result of washing away the pollutants deposited during the preceding dry spell.

Whether the rain will increase or decrease the corrosive action is dependent on the ratio of deposition between the dry and wet contaminants. When the dry period deposition of pollutants is greater than the wet period deposition of surface compounds, the washing effect of the rain will dominate and the corrosive action will be decreased.

In areas where the air is less heavily polluted, the corrosive action of the rain will assume a much greater importance because it will increase the corrosion rate.

Another factor affecting the relative corrosive rate resulting from rain is the orientation of the metal surface. In areas of heavy industrial pollution, skyward-facing metallic surfaces benefit from rain. In those areas where dry deposition is considerably greater than wet deposition of sulfur pollutants, the washing effect of rain predominates, and the corrosion rate is reduced. In areas having less pollution the situation is reversed and the corrosive action of the rain predominates.

C. Fog

In areas having a high degree of air pollution, high acidity and high concentrations of sulfate and nitrate can be found in fog droplets. The pH of fog water has been found to be in the range of 2.2–4.0 in highly contaminated areas.

III. DEPOSIT OF POLLUTANTS

Atmospheric pollutants can be deposited into the aqueous layer by either wet or dry deposition. For wet deposition to take place it is necessary for rain, fog, dew, or snow to be present, whereas for dry deposition precipitation of any kind is not involved. Dry deposition is considered to be predominant indoors or in highly polluted areas close to emission sources. It is difficult to determine the relative importance of wet deposition because of the incidental nature of the precipitation.

Dry deposition is controlled by two factors: aerodynamic processes and surface processes. Aerodynamic processes relate to the actual depletion of the gaseous constituent in the atmosphere (e.g., SO_2), in the atmospheric region adjacent to the aqueous phase, and the ability of the system to add new SO_2 into this region. Whether or not new SO_2 can be introduced into the region is dependent on the actual wind speed, type of wind flow, and shape of the sample. The ability of the surface layer to accommodate the gaseous constituent (e.g., SO_2) is a phenonomen of the aqueous layer, which is a surface process. Greater amounts of gaseous constituents (SO_2) can be accommodated with increases in the thickness of the aqueous layer, resulting from increased relative humidity, the pH of the solution, and the alkalinity of the solid surface.

The dry deposition velocity is defined as the ratio of deposition rate, or surface flux, of any gaseous compound and the concentration of the same compound in the atmosphere. It can be expressed as the inverse of the sum of two resistances, namely, aerodynamic resistance and surface resistance:

$$V_d = \frac{1}{R_a + R_s}$$

where

V_d = dry deposition velocity
R_a = aerodynamic resistance
R_s = surface resistance

In general the dry deposition velocity will be the combined effect of both

Table 3.2 Characteristics of Selected Gaseous Air Constituents[a]

			Eq. conc. (μM)		Deposition velocity (cm/s)		Deposition rate (ng/cm^2)	
Gas	H	M/atm	Outdoor	Indoor	Outdoor	Indoor	Outdoor	Indoor
O_3	1.8	(−2)	2.3 (−4)	1.7 (−4)	0.05–1	0.036	5.8 (−3)	6.8 (−4)
H_2O_2	2.4	(5)	4.2 (3)	1.2 (3)	—	0.07	—	5.0 (−4)
SO_2	1.4	—	1.1 (−2)	2.9 (−3)	0.1–1.2	0.05	7.5 (−3)	2.7 (−4)
H_2S	1.5	(−1)	6.1 (−4)	4.0 (−5)	0.38	0.03	2.2 (−3)	1.1 (−5)
NO_2	7.0	(−3)	1.9 (−4)	3.8 (−5)	0.2–0.8	0.006	2.0 (−2)	6.2 (−5)
HNO_3	9.1	(4)	2.9 (2)	2.7 (2)	0.1–30	0.07	1.4 (−2)	5.5 (−4)
NH_3	1.0	(1)	1.1 (−1)	5.8 (−1)	0.3–2.6	0.05	6.6 (−3)	2.1 (−3)
HCl	2.0	(1)	1.5 (−2)	1.9 (−3)	—	0.04	—	5.8 (−6)
Cl_2	6.2	(−2)	1.2 (−6)	1.4 (−7)	1.8–2.1	—	1.1 (−4)	—
HCHO	1.4	(4)	1.1 (2)	1.4 (2)	—	0.005	—	6.3 (−5)
HCOOH	3.7	(3)	3.3 (1)	7.4 (1)	—	0.006	—	2.3 (−4)

[a]The equilibrium solution concentration and deposition rate values were based on correlations from Table 3.1, and using geometric mean values for the intervals 1.8 (−2) means 1.8×10^{-2}.

Source: Ref. 1.

resistances. However, at highly turbulent air flow conditions $R_a = 0$ and the dry deposition velocity is dependent only on the surface processes. Alkaline surfaces such as lead peroxide or triethanolamine are ideal absorbers of SO_2 for which $R_s = 0$. In this case the dry deposition velocity if dependent on the aerodynamic processes. Typical ranges for dry deposition velocities onto various materials under outdoor and indoor conditions are given in Table 3.2.

In outdoor exposure conditions subject to wet–dry cycles, the actual concentration of most corrosion-stimulating gases under many conditions is not at equilibrium between the gas in the atmosphere and the same gas in the aqueous layer. Even so, thermodynamic considerations have been used for predicting the formation of different corrosion end-products and their stability. Figure 3.1 is a schematic illustration of processes occurring at the aqueous layer.

IV. CORROSION PRODUCTS

One of the most important factors influencing the corrosion rate is the formation and protective ability of the corrosion products formed. The specific corrosion products formed are dependent on the participating dissolved metal ions and the access to anions solved in the aqueous layer. Formation of the film of corrosion products take place in a sequence of consecutive steps—dissolution, coordination, reprecipitation. When the dissolution step is acid-dependent, coordination is based on the hard and soft acid base principle (i.e., hard acids are preferably coordinated with hard bases and soft acids are preferably coordinated with soft bases). Acids or bases with tightly held valence electrons that are not easily distorted are hard acids or bases. Acids or bases having valence electrons that are easily polarized or removed are considered to be soft acids or bases.

Based on experience with atmospheric corrosion, Table 3.3 indicates that hard acids like Cr^{3+} and Ti^{4+} form oxygen-containing films whereas soft acids such as Cu^{+} and Ag^{+} coordinate with reduced sulfur compounds. Intermediate acids such as Fe^{2+}, Cu^{2+}, and Zn^{2+} would be expected to coordinate with a broader range of bases.

The corrosion products formed may be amorphous or crystalline, depending on the rate of crystallization and formation. It is known that slow growth or aging of amorphous phases may result in a change to the crystalline state. This process can occur through slow transformation in the solid state or through dissolution–reprecipitation processes. Such is the case in

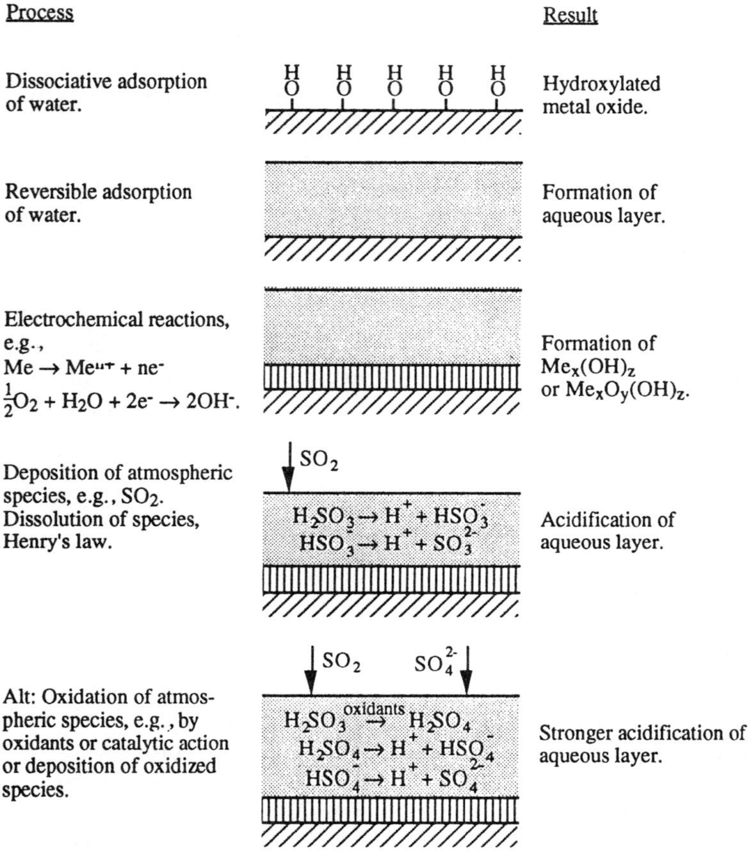

Figure 3.1 Schematic illustration of processes occurring in or at the aqueous layer. (From Ref. 1.)

the transition from amorphous to crystalline state of basic nickel sulfates, with the former being less corrosion-resistant than the latter.

V. SPECIFIC ATMOSPHERIC CORRODENTS

The atmospheric region closest to the earth is known as the troposphere and contains nitrogen, oxygen, and the rare gases Ne, Kr, He, and Xe. Of all the molecules involved these make up 99.9% by weight and of these only oxygen plays a part in atmospheric corrosion. The remaining constituents, ni-

Table 3.3 Classification of Hard and Soft Acids and Bases

Hard	Intermediate	Soft
Acids		
H^+, Na^+, Mn^{2+}, Al^{3+}, Cr^{3+}, Fe^{3+}, Ti^{4+}	Fe^{2+}, Ni^{2+}, Cu^{2+}, Zn^{2+},. Pb^{2+}	Cu^+, Ag^+
Bases		
H_2O, OH^-, O^{2-}, SO_4^{2-}, NO_3^-, CU_3^{2-}	SO_3^{2-}, NO_2^-	R_2S, RSH, RS

trogen and the rare gases, due to their inability to react with metal surfaces are not of significant importance to atmospheric corrosion.

Oxygen, because of its ability to accept electrons and its involvement in chemical transformations of the atmosphere, is particularly important to atmospheric corrosion. Other materials present in the troposphere that play a part in atmospheric corrosion are water and carbon dioxide. Water acts as an electrolyte and carbon dioxide, which has a concentration of approximately 330 ppm and is highly soluble in water, contributes to the acidity of the aqueous layer.

Other trace gases present with a total concentration of less than 10 ppm, which are also of importance in atmospheric corrosion, are O_3, H_2O_2, SO_2, H_2S, COS, NO_2, HNO_3, NH_3, HCl, Cl_2, HCHO, and HCOOH. Either natural or anthropogenic processes are responsible for their presence, and they may undergo a variety of chemical changes during their presence in the troposphere. All species are reactive and exhibit a specific average lifetime, which is limited by their ability to react with important atmospheric oxidizers, specifically the hydroxyl radical OH^- and O_3. The hydroxyl radical is formed by the photoinduced dissociation of ozone and the subsequent reaction of the electronically excited, energy-rich oxygenation O('D) and water vapor:

$$O_3 + hv \rightarrow O('D) + O_2 \quad (\lambda < 310 \text{ nm})$$

$$O('D) + H_2O \rightarrow 2OH^-$$

While it is possible for OH^- molecules to oxidize several of the corrosion-stimulating materials such as SO_2, H_2S, and NO_2, a large portion of the hydroxyl molecules are consumed through reactions with hydrocarbon molecules, producing the hydroperoxyl radical, HO_2, which disproportionates into hydrogen peroxide (H_2O_2) and oxygen (O_2) according to:

$$HO_2 + HO_2 \rightarrow H_2O_2 + O_2$$

Hydrogen peroxide is a powerful oxidizing agent that is highly soluble in water.

A. Sulfur-Containing Compounds

The most important corrosive contaminant found in industrial atmospheres is sulfur dioxide (SO_2), which results from the combustion of sulfur-containing coal and oil, and emission from metal, petrochemical, and pulp and paper industries. Once in the atmosphere the SO_2 undergoes physical and chemical state changes. Depending on the environment, the sulfur dioxide is capable of being oxidized in one or more of the following ways:

$$\text{GM}$$

$$SO_2 + OH \rightarrow HSO_3\bullet$$

$$HSO_3\bullet + O_2 \rightarrow SO_3 + HO_2\bullet$$

$$SO_3 + H_2O \rightarrow H_2SO_4$$

where GM represents another gaseous molecule, such as nitrogen or oxygen, that collides with an $HSO_3\cdot$ molecule and removes some of the excess energy released.

In the aqueous phase:

$$SO_2 + XH_2O = SO_2\bullet XH_2O$$

$$SO_2\bullet XH_2O = HSO_3^- + H_3O^+ + (X - 2)H_2O + H_2O_2$$

$$HSO_3^- \rightarrow SO_4^{2-}$$

On moist particles or in droplets of water the SO_2 may be oxidized to sulfuric acid:

$$SO_2 + H_2O + \tfrac{1}{2}O_2 \rightarrow H_2SO_4$$

Sulfur dioxide has a lifetime in the atmosphere of 0.5–2 days. This limits the distance that the SO_2 may be transported to a few hundred kilometers. During this period the sulfuric acid may be partly neutralized, particularly with ammonia that results from the biological decomposition of organic matter. When this occurs, particles containing ammonium sulfate $(NH_4)_2SO_4$ and different forms of acid ammonium sulfate such as NH_4HSO_4 and $(NH_4)_3H(SO_4)_2$ are formed.

Atmospheric corrosion results from the deposition of these various materials on metal surfaces. Deposition of the sulfur compounds can be accomplished by:

1. Dry Deposition
   a. Adsorption of gas (SO_2) on metal surfaces
   b. Impaction of sulfate particles
2. Wet deposition
   a. Removal of gas from the atmosphere by precipitation in the form of rain or fog.

The primary cause of atmospheric corrosion is dry deposition, which consists mainly of the adsorption of sulfur dioxide. The amount deposited is proportional to the concentration in the atmosphere. Different materials are subject to different deposition rates. Rusty steel surfaces will adsorb SO_2 quantitatively at high relative humidities, whereas the deposition on copper, and particularly aluminum, is much less. The rate of dry deposition of other sulfur compounds is less than that of sulfur dioxide.

Sulfates are deposited primarily by wet deposition and experience a lifetime of 3–5 days. This permits these particles to be transported a distance in the order of 1000 km. Sulfate concentrations are usually less than SO_2 concentrations close to the emission source.

The primary cause of atmospheric corrosion is the dry deposition of sulfur dioxide on metallic surfaces. This type of corrosion is usually confined to areas having a large population, many structures, and severe pollution. Therefore, the atmospheric corrosion caused by sulfur pollutants is usually restricted to an area close to the source.

Under these conditions, dry deposition of SO_2 is considered to be the most important sulfur deposition process for the corrosion of structural metals.

As previously stated, the deposition rate is dependent on the concentration in the air. Since this concentration can vary considerably it is difficult to give ranges. Order of magnitude deposition rates for SO_2 in various types of atmospheres are as follows:

Type of atmosphere	Deposition rate ($mg\ SO_2/m^2$ day)
Rural	<10
Urban	10–100
Industrial	Up to 200

Another atmospheric corrosion stimulant is hydrogen sulfide (H_2S). Natural biological sources such as volcanoes, moss, or swamp areas, and anthropogenic sources such as pulp and paper industries, catalyst converters in motor vehicles, sewage plants, garbage dumps, animal shelters, and geothermal plants, are responsible for the emission of H_2S. Hydrogen sulfide can cause the tarnishing of silver and copper by the formation of tarnish films. Hydrogen sulfide can react with OH^- to form SO_2 as follows:

$$H_2S + OH^- \rightarrow HS\bullet + H_2O$$

$$HS\bullet + 2O_2 \rightarrow HO_2\bullet + SO_2$$

B. Nitrogen-Containing Compounds

High-temperature combustion processes such as in power plants, vehicles, etc., produce NO and NO_2. The combustion gas produced has a much higher percentage of NO than NO_2; however, the NO is rapidly converted to NO_2 according to:

$$2NO + O_2 \rightarrow 2NO_2\bullet$$

At distances further from the emission source the NO may also form NO_2 through the influence of ozone (O_3) as follows:

$$NO + O_3 \rightarrow NO_2 + O_2\bullet$$

The primary nitrogen pollutant near the emission source is nitrogen dioxide (NO_2). The ratio of NO_2 to NO in the atmosphere varies with time and distance from the emission source and is usually between 10 and 100. Nitrogen dioxide can be oxidized to nitric acid according to:

$$NO_2 + OH^- \rightarrow HNO_3\bullet$$

In addition, NO may be oxidized to nitric acid according to the total reaction:

$$2NO + H_2O + \tfrac{3}{2}O_2 \rightarrow 2HNO_3\bullet$$

Since this reaction has a very low rate, the concentrations of HNO_3 and nitrates in the immediate area of the emission source are low.

Nitrogen dioxide, by absorbing solar light and the subsequent formation of ozone through:

$$NO_2 + h\upsilon \rightarrow NO + O \quad (\lambda < 420 \text{ nm})$$

$$O + O_2 \rightarrow O_3$$

plays an important part in atmospheric chemistry.

The mechanisms by which nitrogen compounds are deposited are not completely understood. Wet deposition seems to be the primary mechanism at long distances from the emission source, while in the immediate area of the emission source dry deposition of nitrates appears to dominate. This is due to the fact that NO and NO_2 have a low solubility in water whereas HNO_3, which is highly soluble in water, has not yet formed.

Ammonia (NH_3) is emitted primarily from animal shelters, cleaning detergents, and fertilizer production. Ammonia in the aqueous phase establishes equilibrium with NH_4^-, which results in increased pH. NH_3 affects the atmospheric corrosion chemistry by partly neutralizing acidifying pollutants, forming particulate ammonium sulfate [$(NH_4)_2SO_4$] and acid ammonium sulfates such as NH_4HSO_4 and $(NH_4)_3H(SO_4)_2$.

C. Chlorine-Containing Compounds

In marine environments chloride deposition is in the form of droplets or crystals formed by evaporation of spray that has been carried by wind from the sea. As the distance from the shore increases, this deposition decreases as the droplets and crystals are filtered off when the wind passes through vegetation or when the particles settle by gravity.

Other important sources of chloride emission are coal burning, municipal incinerators, and deicers and dust binders on roads. Most coals have a chlorine content of 0.09–0.15%. Values as high as 0.7% have been found in high-chlorine coals. The combustion of these coals produces an emission of gaseous hydrogen chloride (HCl) that is highly soluble in water and strongly acidifies the aqueous phase.

Many industrial processes, such as bleaching plants in pulp and paper industries, certain metal production facilities, and cleaning detergents, emit chlorine (Cl_2).

Cl_2 can photodissociate into chlorine radicals that react with organic compounds (RH) to form HCl:

$$Cl_2 + h\upsilon \rightarrow Cl\!: + Cl\!: \quad (\lambda < 430 \text{ nm})$$

$$RH + Cl\!: \rightarrow R + HCl\bullet$$

D. Carbon Dioxide (CO_2)

Carbon dioxide occurs naturally in the atmosphere in a concentration of 0.03–0.05% by volume. This concentration varies with the time of day and the season of the year. The above percentages correspond to a concentration of the order of 10^{-5} mol/L when at equilibrium in the water film, if the pH value is 6 or lower.

E. Oxygen (O_2)

Oxygen is a natural constituent of air and is readily absorbed from the air into a water film on the metal surface, which is considered to be saturated thereby promoting any oxidation reactions.

F. Indoor Atmospheric Compounds

HCHO and HCOOH are important indoor corrosion stimulants that can originate from tobacco smoke, combustion of biomass, adhesives, and plastics. In general, the concentration of these stimulants is lower indoors than outdoors, except for ammonia and the organic species, which usually have a higher concentration indoors than outdoors. This higher level is the result of anthropogenic activity.

VI. SUMMARY

The concentrations of pollutants found in both indoor and outdoor atmospheres can vary greatly as a result of the type of atmosphere and/or the geographic location. It is almost impossible to provide a specific range for a specific location unless air samples are taken and analyzed. Listed below are some typical indoor and outdoor ranges of inorganic pollutants as found in the United States.

Pollutant	Outdoor range ($\mu g/m^3$)	Indoor range ($\mu g/m^3$)
SO_2	3–185	1–40
NO_2	20–160	3–60
H_2S	1–36	0.2–1
O_3	10–90	7–65
HCl	0.3–5	0.08–0.3
Cl_2	<5% of HCl levels except where local source exists	0.004–0.015
NH_3	6–12	10–150

REFERENCE

1. Leygraf. Atmospheric corrosion. In: P Marcus, J. Ouday, eds. Corrosion Mechanisms in Theory and Practice. New York: Marcel Dekker, 1995.

4
Resistance to Atmospheric Corrosion of Specific Metals and Alloys

In the previous chapters the factors affecting the mechanisms of atmospheric corrosion have been discussed. However, these factors and mechanisms have different effects and react in different manners with different metals and alloys. Each metal forms its own protective film of corrosion products, some of which offer greater protection than others. These will be discussed in this chapter.

These protective films tend to reduce the corrosion rate with time. Eventually the corrosion rate reaches a steady state and changes very little on further exposure. This is a characteristic of all metals and alloys. The average atmospheric corrosion rate of various metals in mils/year are shown below:

	Atmosphere					
	Urban/industrial		Marine		Rural	
Metal	10 years	20 years	10 years	20 years	10 years	20 years
Aluminum	0.032	0.029	0.028	0.025	0.001	0.003
Copper	0.047	0.054	0.032	0.050	0.023	0.017
Lead	0.017	0.015	0.016	0.021	0.019	0.013
Nickel	0.128	0.144	0.004	0.006	0.006	0.009
Monel	0.053	0.062	0.007	0.006	0.005	0.007
Zinc 99.9%	0.202	0.226	0.063	0.069	0.034	0.044

I. CARBON STEEL

It is logical to start with carbon steel because this is the most widely used structured material. Up to the last 20 years or so, atmospheric corrosion of steel was believed to be a process of general corrosion proceeding in cells with microscopic anodes and cathodes. More recent studies have determined that the corrosion process, which is electrochemical in nature, takes place in cells of microscopic dimensions with very distinct anodic and cathodic areas.

Atmospheric corrosion of steel is a function of location. In country air the products of corrosion are either oxides or carbonates. In industrial atmosphere sulfuric acid is present, and near the ocean some salt is in the air. Corrosion is more rapid in industrial areas because of the presence of the acid, and it is higher both near cities and near the ocean because of the higher electrical conductivity of the rain and the tendency to form soluble chlorides or sulfates, which cause the removal of protective scale.

When steel is exposed in a clean, dry atmosphere, the surface is covered with a 20- to 50-Å-thick oxide film consisting of an inner layer of Fe_2O_3. This film practically prevents further corrosion. If small amounts of water vapor are present FeOOH may also form.

In noncontaminated atmospheres the initiation of corrosion on a clean metal surface is a very slow process, even if the atmosphere is saturated with water vapor. Under these conditions initiation of corrosion may occur at surface inclusions of MnS, which dissolve when the surface becomes wet. However, the presence of solid particles on the surface is a more important factor for the start of corrosion. Airborne dust that has settled on the steel surface may prompt corrosion by absorbing SO_2 and water vapor from the atmosphere. Of greater significance are particles of hygroscopic salts, such as sulfates or chloride, which form a corrosive electrolyte on the surface. Cathodes in microcells may also be formed by carbonaceous particles settling on the steel surface. Cathodes may also be formed by residues of mill scale on the surface, initiating corrosion when the surface becomes wet. Rusting is rapidly initiated in SO_2^--polluted atmospheres at high humidity even in the absence of any particles on the surface.

II. START OF CORROSION PROCESS

Anodic spots surrounded by cathodic areas are created during the initiation period. With an electrolyte film on the metal surface conditions are created for the beginning of the corrosion process, which is further stimulated when SO_2 is adsorbed and oxidized in the rust layer as SO_4^{2-}. Sulfate accumulates at the anodes of the corrosion cell creating so-called nests in the rust. A great number of sulfate nests cover the surface during the initial stage. As exposure continues the number of nests per unit area decreases as the nests grow larger. The type of atmosphere and degree of sheltering will determine the size and distribution of the nests. Tests have indicated that after 4 months of outdoor exposure the average diameter of the nests was approximately 0.5 mm, which increased to approximately 1 mm on further prolonged exposure.

In atmospheres with chlorides present the corrosion of carbon steel proceeds in local cells that resemble the sulfate nests. They may form around chloride particles deposited on the surface, where the concentrated local chloride solution destroys the passivating film of FeOOH. The chlorides are concentrated in the anodic areas formed by migration, while the surrounding area covered by rust acts as a cathode.

The sulfate nests form corrosion cells with the surrounding area when the surface becomes wetted by dew, rain, or moisture adsorption. The electrolyte is very concentrated and has a low water activity. Anodes are present inside the sulfate nests, where the redox potential and pH value become low. These conditions correspond to a position in the Fe^{2+} region in the potential pH diagram, and local attack takes place in the steel surface, with the surrounding area acting as a cathode. This will take place even if the surface is covered with oxide containing crystalline magnetite (Fe_3O_4) because the magnetite is a good electronic conductor. The corrosion reaction is described in terms of an electrochemical cell of the type $Fe/Fe^{2+}(aq)/OH^-/O^2(aq)''Fe_3O_4''$. Hydroxides, as well as magnetite, containing both divalent and trivalent ions (green rust) may serve as cathodes because they possess appreciable electronic conductivity.

The reactions taking place at the corrosion cells may be described by the following equations. The primary cathodic reaction is the reduction of oxygen dissolved in the electrolyte film:

$$\tfrac{1}{2}O_2 + H_2O + 2e^- \rightarrow 2OH^-$$

As a result of this reaction, a local increase in pH takes place at the cathode causing a precipitation of corrosion products at some distance from the anodes.

Once the ferric corrosion products have been formed, another cathodic process may take place:

$$Fe^{3+} + e^- \rightarrow Fe^{2+}$$

The cathodic step then comprises the reduction of ferric rust to magnetite according to:

$$8FeOOH + Fe^{2+}(aq) + 2e^- \rightarrow 3Fe_3O_4 + 4H_2O$$

This process occurs during wet periods.

The anodic basic reaction

$$Fe + Fe^{2+} + 2e^-$$

has different mechanisms in neutral and acid solutions.

When the concentration of sulfates is low the mechanisms of iron dissolution may be described as follows:

$$Fe + H_2O + Fe(OH)^-ads + H^+$$

$$Fe(OH)^-ads \rightarrow Fe(OH)ads + H^+$$

$$Fe(OH)ads \rightarrow Fe(OH)^+ + e^-$$

$$Fe(OH)^+ \rightarrow Fe^{2+} + OH^-$$

When insoluble corrosion products are formed at anodic sites the corrosion rate is reduced due to passivation. This may occur in rural atmospheres with low concentrations of sulfates and chlorides in the rust layer.

However, in sulfate-containing solutions the anodic dissolution proceeds according to the following mechanism:

$$Fe + H_2O \rightarrow Fe(OH)ads + H^+ + e$$

$$Fe(OH)ads + H_2O \rightarrow /Fe(OH)_2/ads + H^+ + e$$

$$/Fe(OH)_2/ads + SO_4^{2+} \rightarrow FeSO_4 + 2OH^-$$

$$FeSO_4 \rightarrow Fe^{2+} + SO_4^{2-}$$

A relatively low pH will result (from the pH-regulating effect of $FeSO_4$) at the anodic sites, thereby preventing precipitation of iron hydroxides directly on the metal surface. This creates favorable conditions for corrosion in the active state, with the sulfate accelerating the anodic dissolution of iron. The crystalline iron at the steel–rust interface is tetrahydrate $FeSO_4 \cdot 4H_2O$. Thus, a reservoir of soluble sulfates exists within the sulfate nests, contributing to their high stability.

In the rust layer

$$2Fe^{2+} + 3H_2O + \tfrac{1}{2}O_2 \rightarrow 2FeOOH + 4H^+$$

This so-called oxidative hydrolysis plays an important role in the majority of the proposed mechanisms of atmospheric corrosion. The magnetite produced by cathodic reduction is reoxidized by oxygen in the presence of water:

$$Fe_3O_4 + \tfrac{1}{4}O_2 + \tfrac{3}{2}H_2O \rightarrow 3FeOOH$$

A semipermeable membrane of hydroxide formed through the oxidative hydrolysis of the iron ions encloses the sulfate nest. The electric current in the corrosion cell causes migration of SO_4^{2-} into the nest, stabilizing the existence of the nest.

In the theory of atmospheric corrosion of steel based on the acid regeneration cycle, sulfuric acid formed by oxidation of SO_2 absorbed in the rust layer attacks the steel according to the overall equation:

$$4H_2SO_4 + 4Fe + 2O_2 \rightarrow 4FeSO_4 + 4H_2O$$

Sulfuric acid is then reformed by oxidative hydrolysis:

$$2FeSO_4 + \tfrac{1}{2}O_2 + 3H_2O \rightarrow 2FeOOH + 2H_2SO_4$$

It should be mentioned that the oxidative hydrolysis of $FeSO_4$ is very slow and the reaction should only affect the corrosion during the initiation stage.

III. SO_2 IN THE CORROSION PROCESS

Sulfur dioxide may be adsorbed on steel surfaces under atmospheric conditions. The rate of adsorption on rusty or polished steel specimens depends on the relative humidity.

Since all sulfur(IV) species in aqueous solutions are thermodynamically unstable, at low and medium concentration of SO_2 (<1–10 ppm) there is a strong driving force for the reaction:

$$SO_2(aq) + H_2O + \tfrac{1}{2}O_2 \rightarrow H_2SO_4$$

The presence of oxides and hydroxides of iron, Fe^{2+}(aq), and Mn^{2+}(aq) catalyzes the reaction, which is strongly pH-dependent, decreasing with pH. At low SO_2 concentrations the pH value will be higher than at high SO_2 levels, thus promoting the oxidation reaction. Sulfur dioxide may be oxidized to the sulfate by Fe^{3+}(aq) and solid ferric oxide hydroxide as well as by oxygen.

High SO_2 concentrations create a low pH in the surface film and consequently a low rate of sulfate production, while the surface film will contain appreciable amounts of tetravalent sulfur. Under these conditions, at SO_2 concentrations greater than 10 ppm, reduction of SO_2 to dithionite may take place according to the following reaction:

$$SO_2(aq) + 2e^- \rightarrow S_2O_4^{2-}$$

The reactive dithionate is then oxidized or further reduced by the metal to the sulfide:

$$S_2O_4^{2-} + 8H^+ + 10e^- \rightarrow 2S^{2-} + 4H_2O$$

Referring to a potential $-$pH diagram (Figure 4.1), it is evident that at high

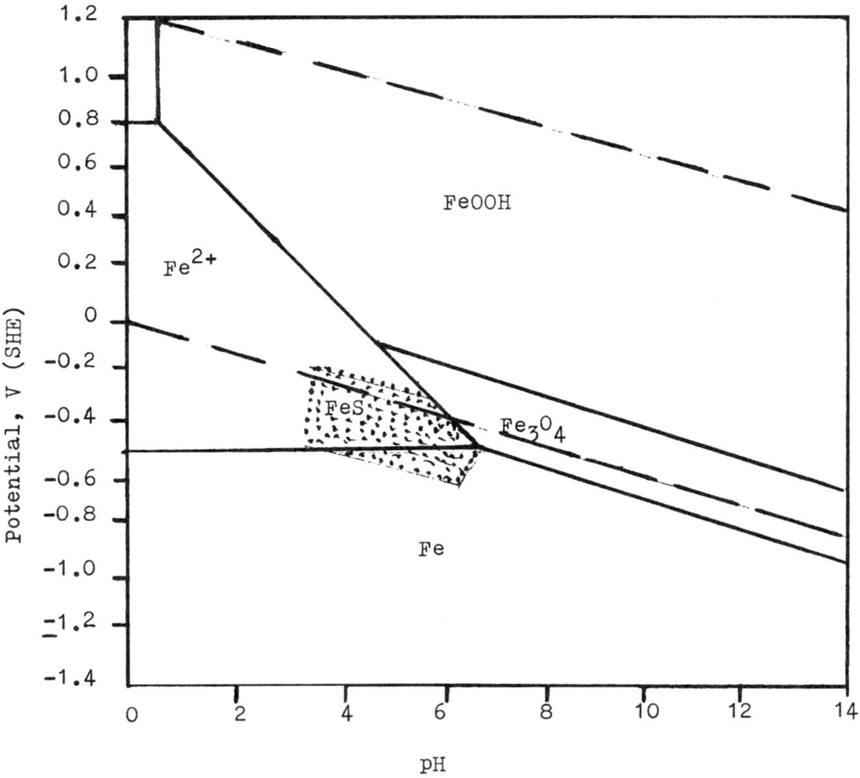

Figure 4.1 Potential–pH diagram, Fe-S-H_2O; 25°C, 1 M Fe, 10^{-2} M S. (From Ref. 1.)

SO_2 concentrations (>10 ppm) FeS is the stable corrosion product in the pH 3–6 range. In a polluted outdoor atmosphere SO_2 concentrations are usually in the range of 0.01–0.2 ppm; therefore, reduction of FeS does not take place. This is particularly true of rusty surfaces, which strongly catalyze the oxidation of SO_2.

The corrosion rate is not constant over a period of time. During the initial period the corrosion rate is usually high because the rust formed is porous and has poor protective properties. This initial period usually lasts 1–5 years, depending on the corrosivity of the atmosphere. After the initial period the protective properties improve and the corrosion rate decreases.

The penetration depth will increase with the time of exposure as shown in Figure 4.2. The penetration curve follows a power law:

$$P = Kt^n$$

where P is the penetration depth, K is a constant, t is the time of exposure, and n is a diffusion constant.

The diffusion coefficient has a constant value of 0.5 when the formation of the rust layer is diffusion-controlled, and all of the rust formed remains on the surface. As the porosity in the rust decreases the value of n decreases to below 0.5. When rust is removed from the surface due to dissolution, flaking, or erosion, the value of n increases to above 0.5. In exposure tests of steel the value of n has been found to vary between 0.4 and 0.8. As the stationary state is reached with constant thickness and porosity of the rust layer, the corrosion rate becomes constant and the relationship between penetration and time becomes linear.

The power law can also be written as

$$\log P = \log K + n \log t$$

which will give a straight line as shown in Figure 4.3.

As discussed in Chapter 2, there are several factors that affect the atmospheric corrosion rate. The most important are the time of wetness and the pollution of the atmosphere, primarily SO_2, and in marine locations chlorides.

In marine atmospheres chlorides play an important role in the acceleration of the corrosion rate of carbon steel. Some industrial areas are also exposed to corrosion by chlorides that originate from the burning of coal.

Oxides of nitrogen appear to have little or no effect on the corrosion rate of carbon steel. This is attributed to the much lower deposition rate of

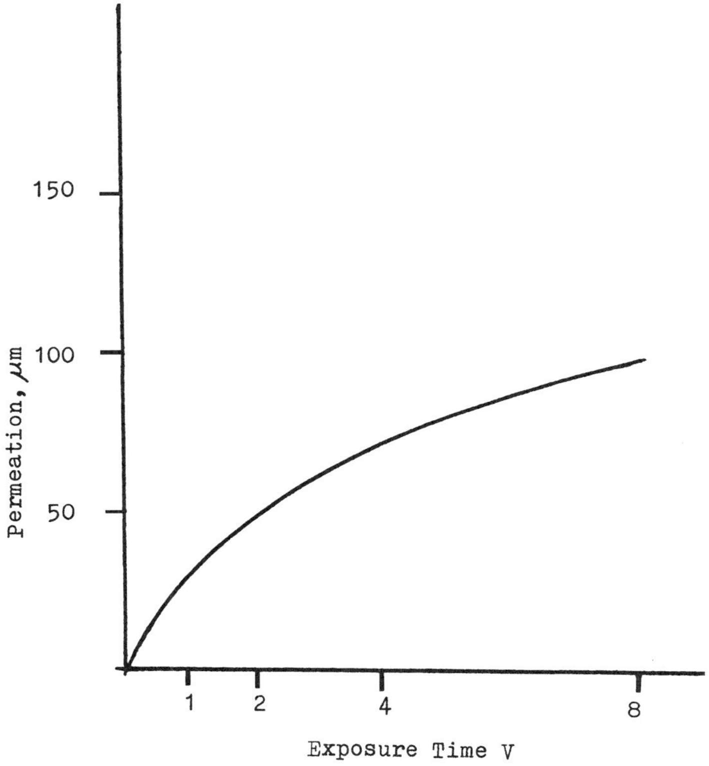

Figure 4.2 Penetration depth vs. exposure time for carbon steel.

NO_x in comparison with SO_2. Oxides of nitrogen have a deposition rate of approximately one-hundredth that of SO_2.

A. Applications

Since carbon steel is not capable of developing its own protective coating, unalloyed steel used in outdoor applications is usually protected with a coating of antirust paint, or zinc or aluminum. In order for the paint coating to be effective, surface preparation prior to painting is extremely important.

The exposure expected of the steel will determine the proper method of surface preparation. Milder environments such as building interiors require a less thorough surface preparation than do surfaces that will be exposed to more severe environments. Table 4.1 provides general surface preparation requirements based on the environment to which the steel will be exposed. See Chapter 14 for specific details.

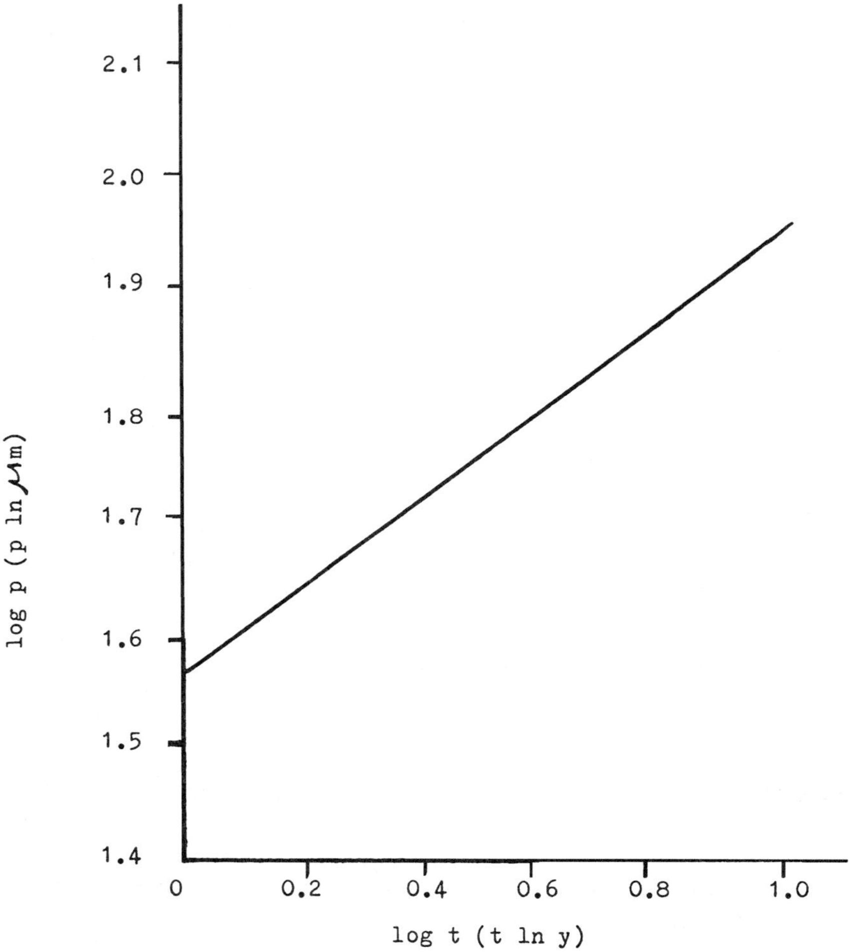

Figure 4.3 Bilogarithmic diagram of penetration depth (*P*) and exposure time (*G*) for carbon steel.

IV. WEATHERING STEEL

Plain carbon steel has little or no resistance against atmospheric corrosion. For this reason a series of low-alloy carbon steels have been developed, referred to as weathering steels. One such example is U.S. Steel's Cor-Ten steel.

The weathering steels are produced by alloying carbon steel with copper, chromium, nickel, phosphorus, silicon, and manganese. These low-alloy

Table 4.1 Surface Preparation Recommendations for Painting

Environment	Recommendation
1. Dry interior: building interior or outside exposure in an arid climate.	Minimal hand or power tool cleaning. Painting for color, aesthetics only.
2. Normally dry but exposed to weather: some bridges, building steel, etc.	Hand or power tool cleaning.
3. Frequently wet or exposure to high humidity: bridges, tanks, topside of ships, steel in paper mills, sewage treatment plants, etc.	Blast clean or pickle.

steels have improved corrosion resistance in outdoor atmospheres, in rural areas, or in areas having relatively low pollution levels. The protective action of copper and other alloying elements is due to a resistant form of oxide that forms a protective coating under atmospheric conditions but has little or no favorable effect when immersed continuously in water.

In an industrial atmosphere, steel with 0.32% copper will corrode only half as much after 5 years as steel with 0.05% copper. A low-alloy, high-strength steel having the following composition (percentage) will corrode only half as much as steel having 0.32% copper:

C	0.12 maximum
Mn	0.20–0.50
P	0.07–0.16
S	0.05 maximum
Si	0.75 maximum
Cu	0.30–0.50
Cr	0.50–1.25
Ni	0.55 maximum

It will be noted that, in addition to copper, this high-strength alloy also contains notable amounts of chromium and nickel, both of which are helpful in increasing strength and adding resistance to corrosion. Phosphorus, which it also contains, is another element that aids in providing protection against atmospheric corrosion. In general, the presence of oxygen or of acidic conditions promotes the corrosion of carbon steel. Alkaline conditions inhibit corrosion. Factors that affect the corrosion resistance of these steels are:

Climatic conditions
Pollution levels
Degree of sheltering from the atmosphere
Specific composition of the steel

Exposure to most atmospheres results in a corrosion rate that becomes stabilized in 3–5 years. Over this period is formed a protective film or patina that is dark brown to violet. The patina is a tightly adhering rust formation on the surface of the steel that cannot be wiped off. Since the formation of this film is dependent on pollution in the air, in rural areas where there is little to no pollution a longer time may be required to form this film. In areas that have a high pollution level of SO_2, loose particles are formed with a much higher corrosion rate. This film of loose particles offers little or no protection against continued corrosion.

When chlorides are present, such as in a marine environment, the protective film will not be formed. Under these conditions corrosion rates of the weathering steels are equivalent to those of unalloyed carbon steel.

In order to form the patina a series of wet and dry periods is required. If the steel is installed in such a manner as to be sheltered from the rain, the dark patina does not form. Instead, a rust lighter in color forms that provides the same resistance. The corrosion rate of the weathering steels will be the same as the corrosion rate of unalloyed steel when it is continuously exposed to wetness such as in soil or water.

Since the patina formed has a pleasant appearance the weathering steels can be used without the application of any protective coating of antirust paint, zinc, or aluminum.

In order to receive the maximum benefit from the weathering steels, consideration must be given to the design. The design should eliminate all possible areas where water, dirt, and corrosion products can accumulate. When pockets are present, the time of wetness increases, which leads to the development of corrosive conditions. The design should make maximum use of exposure to the weather. Sheltering from rain should be avoided.

While the protective film is forming, rusting will proceed at a relatively high rate during which time rusty water is produced. This rusty water may stain masonry, pavements, and the like. Consequently, steps should be taken to prevent detrimental staining effects, such as coloring masonry brown, so that any staining effects will not be obvious. The ground area exposed to staining can be covered with an easily exchangeable material such as gravel.

The corrosion mechanism for weathering steels is similar to that of unalloyed carbon steels. The rust forms a more dense and compact layer on the weathering steels than on unalloyed carbon steels. The rust layer more effectively screens the steel surface from the corrosive environments of the atmosphere. The corrosion process may be affected in several ways by this rust layer. The cathodic reaction may be affected by the low diffusion rate of oxygen, whereas the anodic reaction may be retarded by limiting the supply of water and corrosion-stimulating ions that can reach the surface of the steel. In addition, the increased electrolyte resistance may also decrease the corrosion rate.

Each of the alloying ingredients reacts in a different manner to improve the resistance of the weathering steels to atmospheric corrosion.

Copper has the most pronounced effect of any of the individual elements in decreasing the corrosion rate. An increase in the copper content from 0.01% to 0.4% decreases the corrosion rate by up to 70%. Only a slight improvement in the corrosion resistance results from an increase of copper in the range of 0.2–0.5%. Several theories have been proposed regarding the mechanism by which copper improves the corrosion resistance. One theory is that the beneficial effect is due to the formation of a surface coating of metallic copper, which acts either as protection in itself or promotes anodic passivation by supporting the cathodic reaction. Another theory is that copper ions dissolved from the base metal are able to precipitate sulfide ions originating either from sulfide inclusions in the steel or from the atmospheric pollution, and thus eliminate their detrimental effect. The most probable theory is that copper forms basic sulfates with low solubility, which precipitate within the pores of the rust layer thereby decreasing their porosity. Weathering steels usually contain 0.2–0.5% copper.

When added in combination with copper, chromium and nickel further increase the corrosion resistance of weathering steels. Chromium is usually added in a content of 0.4–1%, whereas nickel is usually added up to 0.65%. Chromium appears to be more effective than nickel. The mechanical properties of the steel are improved by both elements.

Chromium is enriched in the inner rust layer together with copper and phosphorus. They promote the formation of a dense layer of amorphous FeOOH next to the steel surface. This layer acts as a barrier to the transport of water, oxygen, and pollutants. Nickel is supposed to act by forming insoluble basic sulfates in pores of the rust layer.

Phosphorus also helps to improve the corrosion resistance of weathering steels. By increasing the phosphorous content from less than 0.01% to 0.1%, a 20–30% improvement in the corrosion resistance of copper-bearing steels is realized. Phosphorus may form layers of insoluble phosphates in the rust, acting as transportation barriers in the same manner as the basic sulfates previously mentioned. A comparison of the corrosion rates of carbon steel, a copper-phosphorous low-alloy steel, and a chromium-vanadium-copper low-alloy steel are shown in Table 4.2.

Table 4.2 Atmospheric Corrosion of Various Steels in Different Atmospheric Types

		Average reduction in thickness (mil)		
Atmospheric type	Exposure time (yr)	Carbon	A242 (K11510) Cu-P Steel	A558 (K11430) Cr-V-Cu Steel
Urban	3.5	3.3	1.3	1.8
Industrial	7.5	4.1	1.5	2.1
Rural	3.5	2.0	1.1	1.4
	7.5	3.0	1.3	1.5
Severe marine 80 ft	0.5	7.2	2.2	3.8
(25 m) from	2.0	36.0	3.3	12.2
ocean	3.5	57.0	—	28.7
	5.0	[a]	19.4	38.8

[a]Destroy.

A. Applications

Since the rust layer formed is a dark brown patina, which is generally considered attractive, these steels may be used without any additional protective coating of zinc, aluminum, or antirust paint in rural or urban areas. Initially, these steels were used as facades and roofing materials, and gradually found applications in load-bearing structures.

V. ZINC

Zinc is more resistant to atmospheric corrosion than carbon steel. The rate of corrosion of zinc is influenced primarily by the time of wetness and the presence of pollution in the air, specifically CO_2, SO_x, and Cl^-.

Zinc is a relatively base metal. Atmospheric corrosion of zinc starts with the instantaneous formation of a thin film of zinc hydroxide, which may occur in different crystal structures, and the subsequent formation of basic zinc carbonate $Zn_5(CO_3)_2(OH)_6$. The pH of the aqueous layer controls the stability of initial corrosion products and results in the dissolution of Zn^{2+}.

The corrosion process appears to proceed according to the following mechanism (refer to Figure 4.4). Zinc is oxidized to form zinc hydroxide in moist outdoor environments:

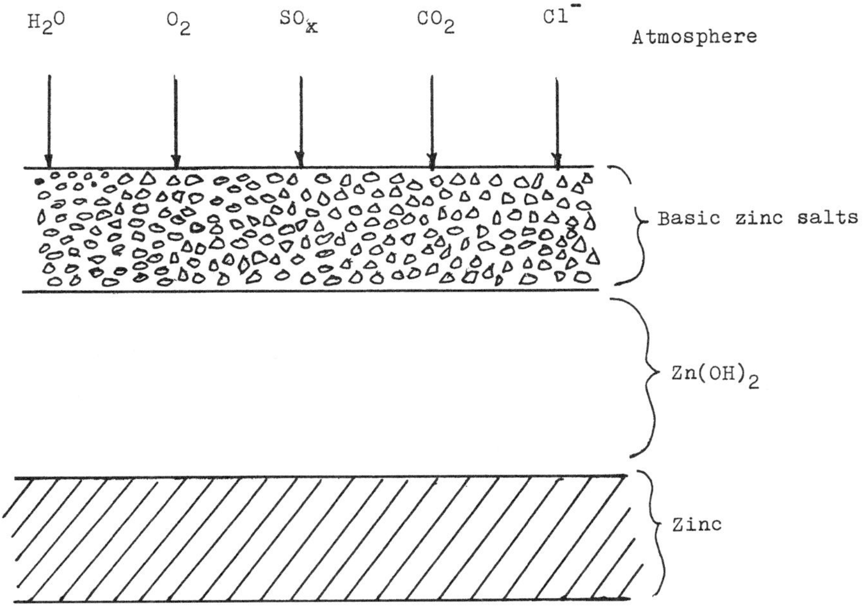

Figure 4.4 Corrosion products formed on zinc during atmospheric exposure. (From Ref. 1.)

$$2Zn + H_2O + \tfrac{1}{2}O_2 \rightarrow 2Zn(OH)_2$$

This reaction being electrochemical involves the anodic oxidation of zinc and the cathodic reduction of oxygen.

If the pH value of the surface moisture is sufficiently high, the zinc hydroxide formed reacts with pollutants present, e.g., CO_2, SO_x, and O_2 to form the corresponding basic zinc salts at the hydroxide–air interface.

$$Zn(OH)_2 + 0.5CO_2 + H^+ \rightarrow ZnOH(CO_3)_{0.5} + H_2O$$

$$Zn(OH)_2 + 0.25SO_2 + 0.25O_2 \rightarrow Zn(OH)_{1.5}(SO_4)_{0.25} + 0.5H_2O$$

$$Zn(OH)_2 + 0.6Cl^- + 0.6H^+ \rightarrow Zn(OH)_{1.4}Cl_{0.6} + 0.6H_2O$$

The zinc hydroxide and the basic zinc salts form the zinc patina which protects the surface from further attack as long as the pH remains high. If due to heavy air pollution with SO_x or from acid rain the pH of the surface moisture is a low value, no zinc hydroxide or basic salt is formed. In addition, any of the basic products previously formed may be dissolved.

$$Zn + SO_2 + O_2 \rightarrow ZnSO_4$$

$$Zn(OH)_2 + SO_2 + \tfrac{1}{2}O_2 \rightarrow ZnSO_4 + H_2O$$

$$ZnOH(CO_3)_{0.5} + SO_2 + O_2 + 2H^+ \rightarrow ZnSO_4 + 1.5H_2O + 0.5CO_2$$

Zinc sulfate is soluble in water and may be washed away by rain leaving the zinc unprotected, producing a high corrosion rate. The protective coating may also be damaged by abrasion and erosion.

As can be seen, the composition of the atmosphere will be a determining factor in the corrosion product formed and will determine whether or not the corrosion product will have protective properties. Table 4.3 illustrates the various corrosion products that may be formed depending on the type of atmosphere present.

Because of this, the prolonged exposure of zinc can follow a number of different paths of reaction sequences, depending on the actual deposition rates of the atmospheric pollutants.

Basic zinc carbonate may continue to grow slowly in a rural atmosphere relatively free of pollutants or it may be followed by the formation of a protective basic zinc sulfate [$Zn_4SO_4(OH)_6 \cdot H_2O$].

When higher amounts of Cl^- than of SO_2 are deposited, as in a marine atmosphere, islands of a less protective basic zinc chloride [$Zn_5Cl_2(OH)_8 \cdot H_2O$] are formed within days of exposure. These islands grow laterally and coalesce. A more protective basic zinc chlorosulfate [$NaZn_4Cl(OH)_6SO_4 \cdot 6H_2O$] may form within weeks of exposure.

When higher amounts of SO_2 than Cl^- are deposited, as in an urban environment, the basic zinc sulfate $Zn_4SO_4(OH)_6 \cdot H_2O$ is formed within weeks of exposure. This is followed eventually by the formation of another

Table 4.3 Potential Zinc Corrosion Products Formed in Various Types of Atmospheres

Atmospheric type	Products formed
Urban or industrial	$Zn \rightarrow Zn(OH)_2 \rightarrow ZnOH(CO_3)_{0.5}$
	$Zn \rightarrow Zn(OH)_2 \rightarrow ZnOH(CO_3)_{0.5} \rightarrow ZnSO_4$
	$Zn \rightarrow Zn(OH)_2 \rightarrow Zn(OH)_{1.5}(SO_4)_{0.25}$
	$Zn \rightarrow Zn(OH)_2 \rightarrow Zn(OH)_{1.5}(SO_4)_{0.25} \rightarrow ZnSO_4$
	$Zn \rightarrow ZnSO_4$
Marine	$Zn \rightarrow Zn(OH)_2 \rightarrow Zn(OH)_{1.4}Cl_{0.6}$
Rural	$Zn \rightarrow Zn(OH)_2 \rightarrow ZnOH(CO_3)_{0.5}$

basic zinc sulfate $[Zn_4Cl_2(OH)_4SO_4 \cdot 5H_2O]$ in highly polluted industrial atmospheres.

In rural and urban atmospheres the penetration depth is generally reported to be an approximately linear function of the exposure time. However, it has been determined that this holds true only for skyward-exposed surfaces. Groundward surfaces in urban atmospheres, as well as groundward and skyward surfaces in marine atmospheres, show nonlinear relations with time in accordance with the power law:

$$p = kt^n$$

where the values of the constants k and n vary with the exposure conditions.

Zinc is also susceptible to a high corrosion rate when so-called white rust is formed in crevices where moisture collects. The white color is due to the formation of zinc carbonate, which does not form a corrosion-limiting protective film on the base metal. Due to lack of protective film formation, white rust corrosion will continue until the protective zinc coating is entirely removed from the underlying steel, which is then subjected to accelerated corrosion and premature failure. White rust will occur whenever zinc or galvanized steel is exposed to water that has a pH value above 8.2. The corrosion rate is governed by the alkalinity of the water and the presence of any accelerating agents. Why the protective basis zinc carbonate cannot form has not been established. It may be due to long-lasting moistening or to a poor supply of CO_2. White rust is more prevalent during rainy seasons. It may also form on moist zinc surfaces in contact with mineral wool, which prevents access of air and drying. This condition may happen in heat-insulated wall structures containing galvanized steel members.

A. Applications

In years gone by zinc coating of steel (galvanizing) was considered a satisfactory corrosion-resistant material for structures in outdoor atmospheres. The corrosion rate was only 0.5–1.0 μm/year. However, the corrosion rate has been increasing in many urban or industrial areas as a result of increasing pollution by SO_2. Corrosion rates have reached 5 μm/year or more in many areas. Because of this it is often necessary to apply to galvanized steel a coat of anticorrosion paint for added protection.

VI. ZINC AND ZINC ALLOY COATINGS ON STEEL

Galvanized steel is steel that has been coated with zinc. The galvanizing process is widely used to protect steel from atmospheric corrosion. Structures, sheet steel, wire, and piping are all forms that are protected by galvanizing.

The protection afforded in rural atmospheres is greater than that in urban or industrial atmospheres. In the latter area there is a greater concentration of industrial pollutants. The air in these areas is contaminated with various sulfur compounds, which together with the moisture in the air convert the normally impervious corrosion-resistant zinc carbonate and zinc oxide layer into zinc sulfate and zinc sulfite. These water-soluble compounds have poor adhesion to the zinc surface and therefore are washed away relatively easily by rain. This exposes the underlying surface to attack by oxygen in the air.

A galvanized steel coating normally corrodes in the same manner as solid zinc. In areas where the zinc coating is defective, the exposed steel, under most conditions, is cathodically protected. Because of the iron-zinc alloy present in most galvanized coatings, galvanized surfaces have a higher corrosion resistance than zinc in neutral and acid conditions. Zinc coatings on steel exposed to air corrode in the same manner as solid zinc, and there is an approximate straight line relationship between weight loss and time. In industrial atmospheres the purity of the coating has little effect on the corrosion rates.

A. Effect of Atmospheric Variables

Protection of galvanized surfaces from atmospheric corrosion is due to the formation of films of basic salts, primarily carbonate. The most widely accepted formula is $Zn_5(OH)_6(CO_3)_2$. The galvanized surface of zinc is rapidly attacked when environmental conditions lead to the formation of soluble films. One of the most important factors affecting the corrosion of zinc in the atmosphere is the duration and frequency of moisture contact.

Since a thin film of moisture with a high oxygen concentration promotes corrosion, the rate of drying is an important factor. Under normal exposure conditions the films dry rapidly, and it is only in sheltered areas that drying times are so slow that the attack of zinc is accelerated.

The effect of atmospheric humidity on the corrosion of zinc is related to the conditions that may cause condensation of moisture on the metal surface and to the frequency and duration of the moisture contact. Moisture is deposited when the air temperature drops below the dew point. The quantity of dew deposited is also influenced by the thickness of zinc, its surface roughness, and its cleanliness. When the temperature of the metal surface falls below the air temperature in a humid atmosphere, moisture condenses on the metal. If the water evaporates quickly, corrosion is not severe and a protective film forms on the surface. If water from rain or snow remains in contact with zinc, when access to air is restricted and the humidity is high, the resulting corrosion can appear to be severe because the formation of a protective basic zinc carbonate is prevented.

Atmospheric corrosion of zinc is roughly proportional to the time of wetness in a particular location provided the nature and quantity of environmental pollution do not change. A high relative humidity, which can cause condensation, increases corrosion.

Time of wetness is normally secondary to the effect of atmospheric pollutants, primarily sulfur oxides and other acid-forming pollutants, which have a direct effect and can also make rain more acidic. In less corrosive areas, time of wetness assumes a greater proportional significance.

B. Effects of Pollutants

Sulfur dioxide is a primary pollutant leading to the atmospheric corrosion of zinc. It controls the corrosion rate when the relative humidity is in the area of 70% or above. Sulfur oxides and other air pollutants are deposited on zinc surfaces either by dry or wet deposition. Regardless of the method of deposition, the sulfur dioxide deposited on the zinc surface forms sulfurous or other strong acids, which react with the protective zinc oxide, hydroxide, or basic carbonate film to form zinc sulfate. The film of protective corrosion products is destroyed by the acids, which reforms from the underlying metal, causing continuous corrosion by an amount equivalent to the film dissolved, hence to the amount of sulfur dioxide absorbed. Corrosion rates increase even further when the relative humidity exceeds 85%.

Atmospheric chlorides will lead to the corrosion of zinc, but to a lesser degree than the corrosion of steel, except in brackish water and flowing seawater. Any salt deposit should be removed by washing. Chlorides also have a much smaller effect on the corrosion rate of zinc than sulfur compounds. The salt content of the atmosphere will usually decrease rapidly inland further away from the coast. Corrosion also decreases with distance from the coast, but the change is more gradual and erratic because chloride is not the primary pollutant affecting zinc corrosion. Chloride is most harmful when combined with acidity resulting from sulfur gases, as would be found on the coast of highly industrialized areas.

Other pollutants also have an effect on the corrosion of galvanized surfaces. Deposits of soot or dust can be detrimental, both because of their specific nature and because they have the potential to increase the risk of condensation onto the surface and hold more water in position. This is prevalent on upward-facing surfaces. Soot (carbon) absorbs large quantities of sulfur, which are released by rainwater.

In rural areas overmanuring of agricultural land tends to increase the ammonia content of the air. The presence of normal atmospheric quantities of ammonia does not accelerate zinc corrosion, and petrochemical plants where ammonium salts are present show no accelerated attack on galvanized steel. However, ammonia will react with atmospheric sulfur oxides, producing ammonium sulfate, which accelerates paint film corrosion as well as zinc corrosion.

When ammonium reacts with NO_x^- compounds in the atmosphere, ammonium nitrite and nitrate are produced. Both compounds increase the rate of zinc corrosion, but less than SO_2 and SO_3.

C. Zinc Alloy Coatings

Combinations of zinc and aluminum are the most important series of alloys. These have been developed for the purpose of improving the atmospheric corrosion resistance of galvanized coatings.

D. Zinc-Aluminum Hot Dip Coatings

The two primary hot dip coatings are Galfan, which is a zinc–5% aluminum alloy, and Galvalume, which is a zinc–55% aluminum–1.5% silicon alloy. Galfan coatings have a corrosion resistance up to three times that of galvanized steel. The main difference between these two coatings lies in the degree of cathodic protection they afford.

Galfan finds application in a mild urban-industrial atmosphere and in a marine atmosphere. This latter application is particularly important because the corrosion rate slows after about 4 years, whereas conventional galvanized steel should show rust in 5 years. This slower rate of corrosion is evidence that the zinc–5% aluminum coatings provide full cathodic protection to cut edges over a longer period of time.

Galvalume provides good protection in urban-industrial atmospheres but at the expense of reduced cathodic protection. The corrosion rate of these coatings also tends to diminish after a year or two.

E. Zinc Aluminum Thermally Sprayed Coatings

Zinc–15% aluminum coatings are available as thermally sprayed coatings. These coatings have a two-phase structure consisting of a zinc-rich and an aluminum-rich phase. The oxidation of the zinc-rich phase takes place within the more inert aluminum-rich phase. The oxidation products formed are encapsulated in the porous layer formed by the latter and do not build up a continuous surface layer as with pure zinc coatings. As a result, no thickness or weight loss is observed even after several years of exposure in atmospheric field testing.

F. Zinc–Iron Alloy Coatings

Compared with pure zinc coatings the zinc-iron alloy coatings provide increased corrosion resistance in acid atmospheres but slightly reduced corrosion resistance in alkaline atmospheres.

VII. CAST ZINC ALLOYS

All of the zinc-aluminum alloys with an 8–29% aluminum content corrode less than 99.99% zinc. ZA-4 (zinc–4% aluminum) castings resist atmospheric corrosion much in the same manner as pure zinc. One advantage is that more zinc may be lost than with zinc coatings without affecting the performance.

In general, corrosion resistance is improved by the addition of higher amounts of aluminum. While ZA-8 (zinc–8% aluminum) and ZA-12 (zinc–12 aluminum) exhibit only modest improvements over the standard 4% alloy, ZA-27 (zinc–27% aluminum) with its significantly larger amount of aluminum-rich phase is much superior.

In plant atmospheres contaminated with either acidic or alkaline fumes the ZA alloys perform better than pure zinc, whereas in alkaline-contaminated atmospheres both zinc and the ZA alloys are more suitable than aluminum.

The addition of copper to zinc alloys provides a slight improvement in corrosion resistance.

VIII. COPPER

For centuries copper has been used as an architectural material. Copper roofs are still in existence on many castles and monumental buildings that are hundreds of centuries old.

When exposed to the atmosphere over long periods of time, copper forms a coloration on the surface known as patina, which in reality is a corrosion product that acts as a protective film against further corrosion. The length of time required to form the patina depends on the atmosphere because the color is due to the formation of copper hydroxide compounds. Initially, the patina has a dark color, which gradually turns green. This is also known as verdigras, which is French for green-gray. In urban or industrial atmospheres the compound is a mixture of copper/hydroxide/sulfate and in marine atmospheres a mixture of copper/hydroxide/chloride. It takes approximately 7 years for these compounds to form. When exposed to clean or rural atmospheres tens of hundreds of years might be required to form the patina, or protective film.

To be classified as copper the compound must contain a minimum of 99.3% copper. Elements such as silver, arsenic, phosphorus, antimony, tellurium, nickel, cadmium, lead, sulfur, zirconium, magnesium, boron, and bismuth may be present singly or in combination.

Copper is noble to hydrogen in the electromotive force (emf) series and thermodynamically stable with no tendency to corrode in water and in non-oxidizing acids free of oxygen. With copper and its alloys the predominant cathodic reaction is the reduction of oxygen to form hydroxide ions. Therefore, the presence of oxygen or other oxidizing agents is necessary for corrosion to take place.

The coating of corrosion products forming the so-called patina has a complex composition that varies from one location to another. Primary components are generally copper oxide and one or more basic copper salts.

Basic sulfate: $Cu(OH)_{1.5}(SO_4)_{0.25}$, $Cu(OH)_{1.33}(SO_4)_{0.33}$
Basic chloride: $Cu(OH)_{1.5}Cl_{0.5}$
Basic carbonate: $CuOH(CO_3)_{0.5}$, $Cu(OH)_{1.33}(CO_3)_{0.33}$, $Cu(OH)_{0.67}(CO_3)_{0.67}$
Basic nitrate: $Cu(OH)_{1.5}(NO_3)_{0.5}$

Basic sulfate is predominant in urban atmospheres and basic chloride in marine atmospheres, whereas in rural atmospheres the primary ingredient is usually basic sulfate. At times basic carbonate is found. The presence of basic copper nitrate indicates that the water film on the metal surface contains an appreciable amount of nitrate.

In most cases the first products generally formed are Cu_2O and CuO. These copper oxides react slowly with pollutants in the air such as SO_x, Cl^-, CO_2, and NO_2, forming the basic salts. For this reaction to proceed, the pH value of the surface moisture must be sufficiently high:

$$Cu_2O + 0.5SO_2 + 1.5H_2O + 0.75O_2 \rightarrow 2Cu(OH)_{1.5}(SO_4)_{0.25}$$

$$Cu_2O + Cl^- + 1.5H_2O + 0.25O_2 \rightarrow 2Cu(OH)_{1.5}Cl_{0.5}$$

Corrosion rates for copper in various atmospheres have been determined as follows:

Atmosphere	Corrosion rate (μm/year)
Rural	1
Urban and industrial	1–3
Marine	1–2

A. Applications

The low corrosion rate and interesting patina formation of copper have been responsible for its use in building structures such as roofs, facades, and gutters. The blue-green patina forms after several years when conditions are favorable in outdoor atmospheres. Initially, it may take on a mottled appearance, but after 6–12 months the surface usually turns to a uniform dark brown color and remains as such for several years. It can take 5–10 years before a green patina begins to appear on sloping surfaces in urban or marine atmospheres. Since vertical surfaces have a much shorter time of wetness they remain black much longer. In a marine atmosphere the surface facing the sea acquires a green patina sooner than the other surfaces as a result of a greater supply of chlorides from the sea winds.

In areas of very low pollution the patina could conceivably take hundreds of years to form because of the small supply of anions to form the basic copper salt. When the temperature is very low or the exposure conditions are very dry, the patina may not form at all.

Under acidic conditions the green patina will not form. Such conditions can occur near chimneys where acid smoke strikes the roof. Under these conditions the corrosion products are soluble.

Traces of dissolved copper will usually be found in rainwater running off of copper-bearing surfaces. This takes place predominately in urban and industrial areas that are highly polluted with SO_x. Such rainwater can cause blue staining on masonry, stonework, etc.

IX. COPPER ALLOYS

To increase the strength and general usefulness of copper, it is alloyed with other elements such as zinc, aluminum, nickel, tin, silicon, and so on. The various alloy families are as follows:

Principal alloying element	Alloy family
Zinc	Brasses
Tin, aluminum, or silicon	Bronzes
Nickel	Copper nickels

There are other alloy families, but these are of little importance in the field of corrosion resistance and will not be discussed.

A. Brasses

Brasses contain zinc as their principal alloying ingredient. Other alloying ingredients are lead, tin, and aluminum. Lead is added to improve machineability and not to improve corrosion resistance. The addition of approximately 1% tin increases the dealloying resistance of the alloys. Aluminum is added to stabilize the protective surface film. Alloys containing in excess of 15% zinc are susceptible to dealloying in acid or alkaline environments.

This is a type of corrosion in which brass dissolves as an alloy and the copper constituent redeposits from solution onto the surface of the brass as a metal in porous form. The zinc constituent may be deposited in place of an insoluble compound or carried away from the brass as a soluble salt. The corrosion can take place uniformly or be local. Uniform corrosion is more apt to take place in acid environments while local corrosion is more apt to take place in alkaline, neutral, or slightly acid environments. The addition of tin or arsenic will inhibit this form of corrosion.

Conditions of the environment that favor dezincification are high temperatures, stagnant solutions, especially of acid, and porous inorganic scale formation. Other factors that stimulate the process are increasing zinc concentrations and the presence of both cuprous and chloride ions. As the dealloying proceeds a porous layer of pure or almost pure copper is left behind. This reaction layer is of poor mechanical strength. The dezincification process on copper-zinc alloys is therefore very detrimental.

High-zinc brasses are also susceptible to season cracking, which is a form of stress corrosion cracking. The term originates from early in the twentieth century when cartridge shells made of 70% copper and 30% zinc were found to crack over a period of time. It was later realized that ammonia from decaying organic matter in combination with residual stresses in the brass was responsible for the cracking of these shells. This phenomenon was called season cracking because the presence of high humidity during warm moist climates (or seasons) promoted the stress corrosion cracking.

Stress corrosion cracking of brass commonly occurs when brass is subjected to an applied or residual tensional stress or while in contact with a trace of ammonia or amine in the presence of moisture and oxygen. The risk of stress corrosion cracking in brasses is greatest in industrial and urban atmospheres, characterized by high contents of sulfur dioxide and ammonia. The stress corrosion susceptibility is markedly lower in marine atmospheres. The relative resistance to stress corrosion cracking of the brasses is as follows:

Low resistance:
Brasses containing >15% zinc

Brasses containing >15% zinc and small amounts of lead, tin, or aluminum
Intermediate resistance:
Brasses containing <15% zinc

The admiralty brasses and the naval brasses are resistant to dealloying as the result of the addition of tin. High-zinc brasses resist sulfides better than low-zinc brasses.

Alloys containing 15% or less zinc resist dealloying and are generally more corrosion-resistant than high-zinc-bearing alloys and are more resistant to stress corrosion cracking than the high-zinc alloys. Red brass is a typical alloy in this group containing 15% zinc. Dissolved air, oxidizing materials, and compounds that form soluble copper complexes (e.g., ammonia) are corrosive to the low-zinc brasses.

B. Bronzes

Copper-tin alloys are known as tin bronzes or phosphor bronzes. Phosphorus is always present in small amounts even though tin is the principal alloying ingredient. Phosphorus is used as an oxidizer and generally has a concentration of less than 0.5%.

These alloys date back to the Bronze Age and probably are the oldest alloys known. Many of the artifacts produced during the Bronze Age are still in existence. Items such as bells, statues, vases, and swords have survived hundreds of years of exposure to a wide variety of environments, proving the corrosion resistance of these materials.

Copper-aluminum alloys are known as aluminum bronzes. They are complex alloys containing iron, nickel, silicon, manganese, and other elements as well as aluminum.

As with copper, the aluminum bronzes find many architectural applications. A special aluminum bronze alloy designated C-61500 has been specifically developed for architectural use. It is reported to have a film resistance 20 times that of copper C-11000. These are suitable for marine atmospheres.

C. Copper-Nickel Alloys

The copper-nickel alloys are referred to as cupronickels and have nickel as the principal alloying ingredient. Of all the copper alloys, the cupronickels are the most resistant to stress corrosion cracking in ammonia and ammonical environments. These alloys do not find many applications to resist atmospheric corrosion because of their cost. They would be most suitable for marine atmospheres.

X. ALUMINUM ALLOYS

The resistance to atmospheric corrosion of aluminum is due to the rapid formation of a thin, compact, and adherent oxide film over the surface that limits corrosion. This surface film when formed in air is approximately 10–20 mm thick. It is thicker when formed in the presence of water or water vapor. This oxide film is stable (insoluble) in the pH range of approximately 4–9, which includes many atmospheric environments.

Pourbaix potential–pH diagrams can be used to predict the regions of oxide stability and of uniform corrosion in atmospheric types as shown in Figure 4.5. Note that aluminum is thermodynamically stable only at low potentials. Because of its property of developing a protective coating of

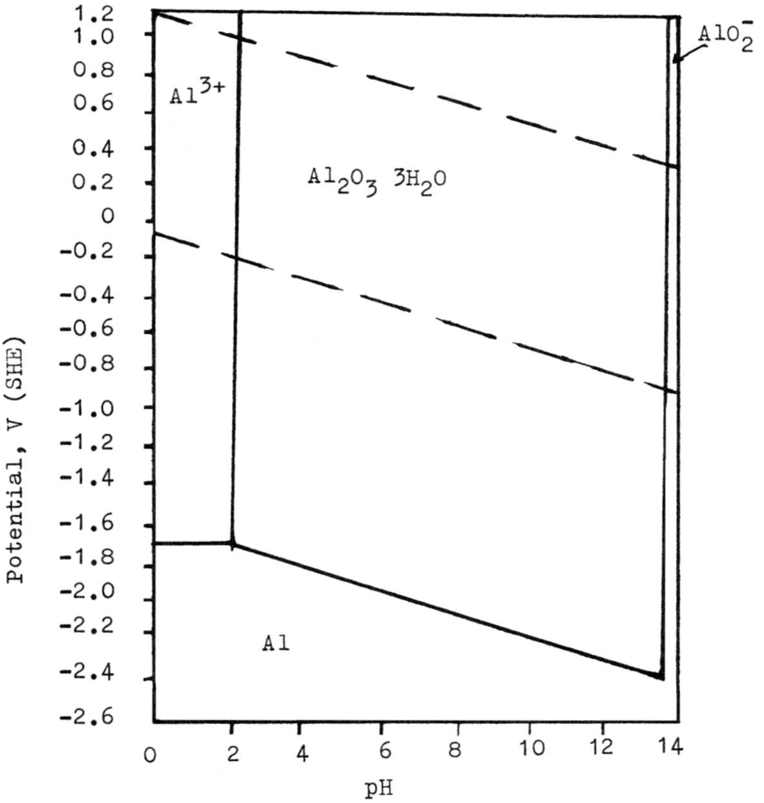

Figure 4.5 Potential–pH diagram, Al-H_2O; 25°C, 10^{-1} M Al. (From Ref. 1.)

alumina it can be used in the presence of water. At an Al^{3+} activity of 10^{-1} μ the stability range of the oxide coating extends down to pH 2.5.

Corrosion products that form during the outdoor atmospheric corrosion of aluminum are amorphous Al(OH)3, α-$Al(OH)_3$ (bayerite), and γ-Al_2O_3, the latter with varying amounts of water in the lattice.

A 10- to 20-mm-thick amorphous oxide protective coating forms when a fresh aluminum surface is exposed to dry air. In a moist atmosphere the oxide coating will thicken. The coating consists of one dense, protective barrier layer next to the metal and an outer, more permeable bulk layer.

Water-soluble salts such as $Al_2(SO_4)_3$ may be formed when anions (e.g., SO_4^{2-} or Cl^-) are deposited on the oxide surface and react with the oxide film. They may also be incorporated into the lattice to form a variety of basic salts and complexes.

In urban atmospheres the oxide coating is protective against SO_2 pollution, producing a relatively low pH value in the moisture film. If high SO_2 pollution exists a very low pH value of the moisture film is produced, which leads to dissolution of the protective film.

In the presence of chloride ions, pitting may be initiated. It is believed that the chloride ions migrate to the oxide layer and lower its resistance to outward migration of Al^{3+}. During the propagation stage aluminum is dissolved anodically to Al^{3+} ions within the pit. A cathodic reaction takes place either inside the pit, close to its mouth, or outside the pit and consists of the reduction of oxygen or H^+ ions. The passivating oxide layer has low electronic conductivity, but the cathode reaction may locally destroy the protective oxide layer due to alkalization. This lowers the electrode potential and may even make hydrogen liberation possible. Acid conditions are created within the pit by hydrolysis of the Al^{3+} ions, and a cap of $Al(OH)_3$ and/or Al_2O_3 forms over the mouth of the pit. The corrosion products finally block the operation of the pit.

A. Atmospheric Weathering

Aluminum retains its shiny appearance for years in clean outdoor atmospheres. Small pits, which are barely visible to the naked eye, develop in polluted outdoor atmospheres. These pits become covered with crusts of aluminum oxide and aluminum hydroxide. During the first few years of exposure the growth rate of the maximum pit depth is relatively high. This growth rate decreases gradually, so that the pit depth approaches a nearly constant value (see Figure 4.6).

The corrosion resistance of aluminum is reduced by most alloying constituents in clean or slightly polluted atmospheres.

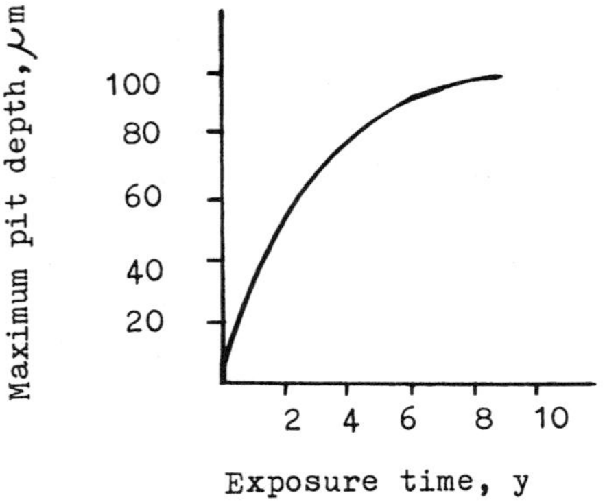

Figure 4.6 Maximum pit depth vs. exposure time for Al MN 1.2 in an urban atmosphere. (From Ref. 1.)

B. Urban or Industrial Atmospheres

Urban or industrial atmospheres are those atmospheres that are high in CO and CO_2, sulfates and sulfites, and possibly various N_xO gases. The corrosion rate is largely the same for unalloyed aluminum as with most conventional aluminum alloys (alloys with Zn, Mg, Si, and/or Mn). Aluminum alloys with copper will exhibit corrosion rates 4–20 times higher. When the SO_2 concentration is less than 0.01% by volume, the atmospheric corrosion rate, even at a relative humidity of 98%, is negligibly affected by the presence of the SO_2. This is probably due to the low adsorption tendency for SO_2 on aluminum surfaces. When the SO_2 concentration exceeds 0.01% by volume, severe corrosion effects occur.

Some highly urban areas, such as the greater Chicago area, have been found to cause exfoliation, whereas other cities such as Los Angeles, which have high N_xO gases, are capable of causing increased susceptibility to stress corrosion cracking. However, in general, most urban/industrial areas do not cause exfoliation and the need for protection is to:

1. Retain aesthetic appearance
2. Prevent pitting formation of their parts
3. Avoid failure by stress corrosion cracking

The aesthetic problem of discoloration can be caused by excessive smoke emissions. This was a major problem before regulations on smoke emissions were enacted. The discoloration problem is not limited to aluminum, but also occurs on other metals, masonry, and stone.

Practically all of the aluminum alloys, at least with appropriate protective measures, will provide satisfactory corrosion resistance in urban/industrial areas. Consequently, the choice of alloy can be based on other engineering considerations.

C. Seacoast Atmospheres

The most damaging natural atmosphere for aluminum is that of a seacoast, with the severest corrosion taking place within 0.5 mile (0.8 km) of the shore line. Corrosive effects will be greatly reduced at distances of 2–5 miles (3–8 km) from the shore. Prevailing wind directions and the roughness of the surf determine the extent of the corrosive effect, which controls how far inland actual salt mist will be carried. In order to maintain the shining metallic surface of aluminum in a seacoast atmosphere, pure aluminum as well as some aluminum alloys must be protected with a coat of clear lacquer or an anodized coating.

Alloys of the 2XXX and 7XXX with tempers that are susceptible to intergranular corrosion, exfoliation, or stress corrosion cracking should not be used in seacoast atmospheres. Thin products should be afforded protection against perforation even if they are only susceptible to pitting corrosion. For alloys whose tempers are resistant to exfoliation and stress corrosion cracking the need for protection is dependent on two factors;

1. The desire to retain a pleasing appearance.
2. Whether the application is fatigue-critical. If so, the depth of pitting corrosion might cause sufficient stress to initiate a fatigue crack.

Structural products about 12 mm or more in thickness that are used for non-fatigue-critical structural applications will darken and have the surface stained as a result of weathering. Pitting corrosion will roughen the surface to a maximum depth of approximately 0.15 mm. However, the loss in residual tensile strength and load-carrying ability will be negligible.

The higher strength alloys such as 5XXX and 6XXX do not require any protective coating since they have sufficient inherent corrosion resistance. These alloys will darken and their surface will undergo a slight roughening, but they are capable of surviving 20 or more years of exposure without appreciable degradation.

D. Rural Atmospheres

Until recently rural atmospheres were considered as noncorrosive to aluminum alloys. However, airborne pollution from coal-burning power plants located upwind have produced an acid rain that falls over a widespread area. This has resulted in staining and darkening of the aluminum surfaces but has not resulted in more severe forms of corrosion. Acid rainfalls in rural areas have caused greater corrosion problems on steel than on aluminum.

XI. PROTECTIVE MEASURES

Proper design of the product along with the selection of the proper alloy are the first steps that can be taken toward protection of the finished product. When necessary there are various coatings that can be applied.

A. Anodic Coatings

It is possible to apply an anodic coating to aluminum alloy products. The coating can be a spray coating of zinc, pure aluminum, or a more anodic aluminum alloy.

B. Alclad Coating

Alclad coating consists of a metallurgical bond between two pure aluminum sheets forming a sandwich on either side of an aluminum alloy sheet that is referred to as the core. The cladding process provides a barrier layer of a more resistant material than the core. Should the barrier layer become damaged by inadvertent scratches and dings, or accidentally penetrated by drilling or sawing, or if natural pitting corrosion of the cladding reaches the core, the electrochemical nature of the cladding takes over. Any further corrosion spreads laterally, confining itself to the cladding rather than penetrating to the core. Cladding is one of the few surface protection methods that provides effective protection even when a break or flaw exists in the coating.

C. Barrier Coatings

Barrier coatings are of two types. The first consists of a thin surface layer of a more resistant metal. This can be accomplished by roll bonding (e.g., stainless steel to non-heat-treatable aluminum alloys), by chrome plating, or by vapor deposit or ion deposition of a variety of metals. The latter methods are expensive and have size limitations. Unlike Alclad coatings, metal barrier coatings do not provide any electrochemical protection at a break in the coating.

Conversion coatings and organic paint coatings are the more conventional barrier coatings used (see Chapters 6 and 7).

D. Anodized Coatings

Anodized coatings are produced by an electrolyte process. The surface of the aluminum is immersed in an acidic bath and made the anode, and a direct current is applied. Oxidation of the surface produces a hard porous film of aluminum oxide that is bound as tenaciously to the alloy as the natural oxide film but is much thicker. Most anodized coatings are produced using either sulfuric acid, chromic acid, or phosphoric acid baths. Other solutions can also be used (e.g., boric acid, oxalic acid). The porous oxide film is sealed by immersion in boiling water, which imparts impermeability to the film. The porous film may be dyed to impart color before sealing.

The anodized surfaces are shinier and more wear-resistant than the natural oxide surface and are easier to keep clean. Eventually they will wear or roughen through use. The primary attributes of anodizing are its decorative appearance and ease of cleaning.

Anodized coatings are not resistant in most environments having a pH outside the range of 4–9. Within this range the corrosion resistance is improved by an order of magnitude or more. Atmospheric weathering tests show that the number of pits that developed in the base metal decrease exponentially with the thickness of the coating. As a general rule, anodized coatings do not provide protection sufficient to make aluminum alloys suitable for environments where, without the coatings, the alloys themselves are unsuitable because of their poor resistance in them.

Anodized surfaces are excellent for subsequent painting but usually less expensive chemical treatments are used for this purpose (see also Chapter 6).

E. Paints

Aluminum is painted for both protective and decorative purposes. As with any metal, surface preparation is extremely important. Conversion coatings, either of the chromate or phosphate type, are recommended for the preparation of aluminum alloys. In environments containing chlorides a chromated primer should be applied first. Water base paints are particularly suitable for use with aluminum alloys (see Chapter 7).

F. Applications

Aluminum and aluminum alloys find applications in outdoor atmospheres as roofing, building facades, and window frames. They are particularly useful in SO_2-polluted atmospheres because of their relatively good resistance. After decades of exposure aluminum structures in urban atmospheres have been reported to be in excellent condition.

The shiny metal appearance will gradually disappear and the surface will roughen under the formation of a gray patina of corrosion products. If soot is present in the atmosphere, it will be absorbed by the corrosion products and give the patina a dark color. If the surface has been anodized, the shiny metal appearance will be retained. Except in highly polluted atmospheres the low penetration rate and shallow pitting do not usually affect the mechanical properties of the aluminum.

Dust and other pollutants may collect in rain sheltered positions and accelerate corrosion by disturbing the formation of a protective oxide coating. These rain-sheltered areas will often show staining as a result of the corrosion. When possible, such rain sheltered areas should be eliminated during the design stage.

Aluminum and its alloys are also suitable for use in marine atmospheres. Care must be taken in the design since under these conditions the aluminum is susceptible to bimetallic corrosion when in contact with copper or carbon steel. If the chloride concentration is high, the aluminum may be susceptible to crevice corrosion.

XII. STAINLESS STEEL

Stainless steel is probably the most widely known material that is used for corrosion resistance. However, stainless steel is not a singular material as the name might imply, but rather a broad group of alloys each of which exhibits its own physical and corrosion resistance properties.

Stainless steels are alloys of iron to which a minimum of 11% chromium has been added to provide a passive film to resist rusting when the material is exposed to weather. This film is self-forming and self-healing in environments where the stainless steel is resistant. Improved corrosion resistance is achieved as more chromium is added to the alloy. Consequently, there are stainless steels with chromium contents of 15%, 17%, 20%, and higher. Other alloying ingredients are added to further improve the corrosion resistance and mechanical strength.

As a result of these alloying possibilities, more than 70 stainless steels are available. They can be divided into four major categories depending on their microstructure:

Austenitic
Ferritic
Martensitic
Duplex

Resistance to atmospheric corrosion was one of the earliest characteristics of the chromium-bearing alloys. When compared to iron and steel, their resistance prompted the name "stainless."

Although atmospheric corrosion of ferrous metals is related first to the amount of moisture in the air (relative humidity), atmospheric moisture alone has no influence on the corrosion of stainless steel. Of primary importance are the effects of such atmospheric contaminants as particulate chlorides and iron-based dust. Sulfur-based acids will promote corrosion while oxides of nitrogen improve the resistance. Contaminants such as hydrocarbons, ammonia, and oxides of carbon have no effect.

A. Austenitic Grades

Of all the stainless steels the austenitic grades are the most widely used for corrosion resistance. By adding substantial quantities of nickel to high-chromium alloys the austenite is stabilized at room temperature. This group of alloys contains 16–26% chromium and 6–22% nickel while the carbon content is kept low (0.08%) to minimize carbide precipitation. The types of stainless steels that make up the austenitic group are as follows:

201	304N	316SF	20Cb3
202	305N	316L	904L
205	308	316N	20Mo4
301	309	317L	20Mo6
302	309S	317LN	Al-6XN
202B	310N	321	25-6Mo
303	310S	330	Alloy 31
304	314S	347	Alloy 654
304L	316	348	

The corrosion resistance of the austenitic stainless steels is the result of the formation of an oxide film on the surface of metal. Consequently, they perform best under oxidizing conditions. Reducing conditions and chloride ions destroy the film causing rapid attack.

Types 304 and 304L stainless steels are normally used in atmospheric exposures. In industrial or marine environments types 316 and 316L are usually selected because of their greater resistance to chloride contamination as the result of the addition of molybdenum.

Any of the austenitic stainless steels may be used in rural or other uncontaminated atmospheres. Under these conditions they will serve indefinitely without significant changes in appearance or losses in strength, even if the relative humidity approaches 100%. Selections for such applications would be based on cost, availability in the sizes and shapes required, mechanical properties, ease of fabrication, and appearance.

In industrial or marine atmospheres, or from contamination with highway deicing salts, chloride pitting may occur. If the contamination is severe, overall rusting may occur rather than localized pitting.

Design of the installation in an industrial atmosphere is important. If the stainless steel is partially sheltered so that deposits are not washed away by rain, and the deposits are hygroscopic and corrosive, some attack may take place that should not otherwise be encountered. In tests, exposed specimens were virtually unattacked, whereas the specimens in a sheltered location under a roof, only those containing molybdenum (types 316 and 316L) remained unattacked.

In marine atmospheres types 304 and 304L stainless steels may develop superficial staining in the form of scattered patches of yellow-brown films. There is little evidence of attack beneath the films. The discoloration will develop during the first few months of exposure after which it does not appear to develop further. Even after exposure of 10 years or more the stain may easily be removed revealing a bright surface with little evidence of attack. If the staining is unacceptable a more highly alloyed stainless such as type 309 will evidence a greatly reduced amount of staining. The use of type 316 will practically eliminate all staining.

As the distance from the shore line increases the intensity of the staining will be reduced. However, the type 316 alloy remains free from staining for long periods of time even when exposed as close as 80 ft from the ocean.

Any of the austenitic stainless steels can be used in a marine atmosphere if they are cleaned periodically. If cleaning is not practical, then type 316 alloy should be used because this alloy will provide the greatest resistance to staining.

B. Ferritic Grades

The ferritic stainless steels contain 15–30% chromium with a low carbon content (0.1%). Resistance to atmospheric corrosion of the ferritic grades will depend on the chromium content as well as the condition of exposure.

Types 405 and 409 have approximately the same resistance to chloride pitting as type 410, whereas higher alloyed types such as 430 are more resistant.

Rural atmospheres are less corrosive to the ferritic grades than marine atmospheres, while the corrosiveness of industrial atmospheres depends on the nature of the contaminants.

Type 430 will rust in a sheltered location when exposed to an industrial atmosphere. If exposed to washing by rain the rusting can be reduced or eliminated.

C. Martensitic Grades

The martensitic stainless steels normally contain 11–13% chromium. Martensite is a body-centered tetragonal structure that provides increased strength and hardness vs. the annealed stainlesses with other lattice structures. Sufficient carbon is added to permit martensite formation with rapid cooling. Other elements such as nickel or molybdenum may be added for improved corrosion resistance.

The corrosion resistance of the martensitic stainless steels is inferior to that of the austenitics. These alloys are usually used in mildly corrosive applications such as atmospheric.

Alloy 17-7PH is a semi-austenitic stainless steel. In annealed condition it is austenitic, but in the aged condition it is martensitic.

In rural and mild industrial atmospheres it has excellent resistance to general corrosion, being equivalent to type 304 stainless steel. In marine atmospheres alloy 17-7PH will gradually develop a light rusting and pitting in all heat-treated conditions. It is almost equal to type 304 in these applications.

D. Duplex Grades

The duplex stainless steels are alloys whose microstructures are a mixture of austenite and ferrite. Type 32550 and related alloys containing molybdenum have an atmospheric corrosion resistance superior to type 316 in high-chloride environments. The duplex grade would be used when the greater strength of these alloys is required for application in a marine atmosphere.

XIII. NICKEL

Nickel 200 outdoors becomes dull and acquires a thin adherent corrosion film that is usually a sulfate. The rate of corrosion is extremely slow; however, the rate increases with an increase in the sulfur dioxide content of the atmosphere. Nickel 200 has a very low corrosion rate in both rural and marine atmospheres.

A. Monel Alloy

Monel 400 is a nickel alloy containing 30–33% copper. The corrosion of Monel 400 is negligible in all types of atmospheres. When exposed to rain, a thin gray-green patina develops. In a sulfurous atmosphere a smooth, brown adherent film forms. Because of its low corrosion rate and the pleasing patina that forms, Monel 400 is used for architectural applications such as roofs, gutters, and flashings.

B. Inconel Alloy 600

Inconel 600 is a nickel-chromium-iron alloy having approximately 16.0% chromium and 7.0% iron. In rural atmospheres the alloy remains bright for many years. When exposed to an industrial atmosphere containing sulfur a slight tarnish may develop. It is preferable for Inconel 600 to be exposed to the atmosphere rather than be sheltered. This will permit rain to wash the surface, and the sun and wind to dry the surface.

C. Inconel Alloy 625, Incoloy Alloy 800, Incoloy Alloy 825

Inconel 625 is a nickel-chromium-molybdenum-columbium alloy. Incoloy 800 is a chromium-nickel-iron alloy. Incoloy 825 is a chromium-nickel-iron alloy similar to alloy 800 but having a higher nickel content. These alloys remain bright and tarnish-free in rural and industrial atmospheres. Alloy 800 develops a very thin oxide film after prolonged exposure in a marine atmosphere, whereas alloys 625 and 825 remain bright.

XIV. TEMPORARY PROTECTION WITH INHIBITORS

Occasions arise when temporary protection of metallic surfaces against atmospheric corrosion is required. Typical instances are the case of finished metallic materials or of machinery parts during transportation and/or storage prior to use. When ready to be used the surface treatment or protective layer can be easily removed.

It is also possible to provide protection by controlling the aggressive environment either by eliminating the moisture and the aggressive gases or by introducing a vapor phase inhibitor. This latter procedure can only be accomplished in a closed environment such as a sealed container, internal car parts, and museum showcases. Treatment with vapor phase inhibitors offers the advantage of preserving the objects independent of their form and their surface state.

Organic substances used as contact inhibitors are compounds belonging to the following classes:

1. Aliphatic, cycloaliphatic, aromatic, and heterocyclic amines.
2. Amine salts with carbonic, carbamic, acetic, benzoic, nitrous, and chromic salts.
3. Organic esters
4. Nitro derivatives
5. Acetylenic alcohols

REFERENCES

1. V Kucera, E Mattson, Atmospheric Corrosion. In: F Mansfield, ed. Corrosion Mechanisms. New York: Marcel Dekker, 1989, pp 211–284.
2. FC Porter, Corrosion Resistance of Zinc and Zinc Alloys. New York: Marcel Dekker, 1994.

5
Atmospheric Corrosion of Metallic Coatings

The purpose of a coating is to extend the time required before corrosion attacks the substrate. Properties of the coating material and the severity of the corrosive atmosphere will determine the service life of the coating. These factors must be taken into account when selecting the coating material.

The corrosion loss of a coating metal in the atmosphere is given by the following equation:

$$L = VKT \tag{5.1}$$

where

L = corrosion loss of metal (mil)
V = average corrosion rate of coating during the wet period (mil/year)
K = ratio of time of wetness for 1 year
T = exposure time (years)

The concentration of aggressive pollutants determines V, whereas K is determined by the location of the coated material.

In order for a coating to be effective in preventing corrosion it must be capable of isolating the base metal from the atmosphere. The ability of the coating to perform this function is dependent on the physical quantity (thickness of coating layer) and chemical properties of the coating. The chemical properties are responsible for the durability of the coating material in a specific environment, i.e., the corrosion resistance of the metal coating.

The corrosion resistance of the coating material must be greater than that of the base metal or it must be maintained by some means. At times it may also be required of the coating material to protect the base metal from attack when the coating material has a defect resulting from mechanical damage and/or pitting corrosion.

The most commonly used corrosion-resistant metallic coatings are classified into five categories based on electrochemical principles:

1. Electromotive force (EMF) control protection
2. Cathodic control protection
3. Anodic control protection
4. Mixed control protection
5. Resistance control protection

The mechanism of the corrosion cell explains the theories of these protective abilities.

I. CORROSION CELL

When a metal is exposed to an atmosphere in which water and oxygen are present, a corrosion cell forms on the metal surface. The electrochemical reactions taking place in a corrosion cell can be written in the following general forms, where (M is a metal):

Anodic reaction:

$$M \rightarrow M^{n+} + ne \tag{5.2}$$

Cathodic reaction in acidic solution:

$$2H^{+} + 2e \rightarrow H_2 \tag{5.3}$$

In neutral and alkaline solutions:

$$O_2 + 2H_2 + 4e \rightarrow 4OH^{-} \tag{5.4}$$

The Evans diagram shown in Figure 5.1 represents the mechanism of the corrosion cell. The cathodic current is expressed in the same direction as the anodic current.

In Figure 5.1, E_c shows the single potential for H_2/H^+ or for O_2/OH^- at the cathode and E_a shows the single potential for metal/metal ion equilibria at the anode. The single potential is given by the Nernst equation:

$$E = E^0 + \frac{RT}{nF} \ln a \tag{5.5}$$

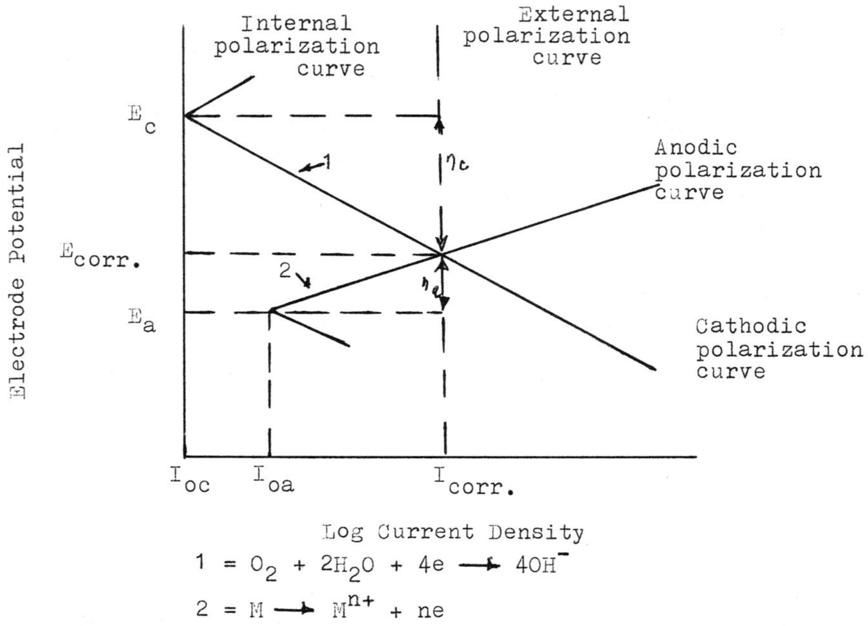

Figure 5.1 Mechanism of corrosion cell. (From Ref. 1.)

where

E = single potential
E^0 = standard single potential
R = absolute temperature
n = charge of an ion
F = Faraday constant
a = activity of the ion

When $a = 1$, E is equal to E^0. The standard single potential E^0 shows the degree of activity of metal and gas.

The electrochemical series consists of the arrangement of metals in order of electrode potential. The more negative the single potential is, the more active the metal. Table 5.1 provides the standard single potential of various metals and nonmetal reactants.

When the electromotive force ($E_c - E_a$) is supplied, the corrosion cell is formed with the current flowing between the anode and the cathode. The cathodic electrode potential is shifted to the less noble direction and the anodic electrode potential is shifted to the more noble direction. The shifting of potentials are called cathodic and anodic polarization. The reaction rate

Table 5.1 Standard Single Potentials, E^0 (V, SHE, 25°C)

Active		Inert	
Electrode	E^0	Electrode	E^0
Li/Li^+	−3.01	Mo/Mo^{3+}	
Rb/Rb^+	−2.98	Sn/Sn^{2+}	−0.140
Cs/Cs^+	−2.92	Pb/Pb^{2+}	−0.126
K/K^+	−2.92	H_2/H^+	±0
Ba/Ba^{2+}	−2.92	Bi/BiO	+0.32
Sr/Sr^{2+}	−2.89	Cu/Cu^{2+}	+0.34
Ca/Ca^+	−2.84	Cu/Cu^+	+0.52
Na/Na^+	−2.71	Rh/Rh^{2+}	+0.6
Mg/Mg^{2+}	−2.38	Hg/Hg^+	+0.798
Th/Th^{4+}	−2.10	Ag/Ag^+	+0.799
Ti/Ti^{2+}	−1.75	Pd/Pd^{2+}	+0.83
Be/Be^{2+}	−1.70	Ir/Ir^{3+}	+1.0
Al/Al^{3+}	−1.66	Pt/Pt^{2+}	+1.2
V/V^{2+}	−1.5	Au/Au^{3+}	+1.42
Mn/Mn^{2+}	−1.05	Au/Au^+	+1.7
Zn/Zn^{2+}	−0.763	O_2/OH	+0.401
Cr/Cr^{3+}	−0.71	I_2/I^-	+0.536
Fe/Fe^{2+}	−0.44	Br_2/Br^-	+1.066
Cd/Cd^{2+}	−0.402	Cl_2/Cl^-	+1.356
In/In^{3+}	−0.34	F_2/F^-	+2.85
Ti/Ti^+	−0.335	S/S^{2-}	−0.51
Co/Co^{2+}	−0.27	Se/Se^{2+}	−0.78
Ni/Ni^{2+}	−0.23	Te/Te^{2+}	−0.92

Source: Ref. 1.

curves ($E - i$) are known as cathodic or anodic polarization curves. The corrosion potential (E_{corr}) and the corrosion current (i_{corr}) are given by the intersection of the cathodic and anodic polarization curves—indication that both electrodes react at the same rate in the corrosion process.

The polarization curves in the current density of more than i_{corr} are called external polarization curves and those in the current density range of greater than i_{corr} are called internal polarization curves. By sweeping the electrode potential from the corrosion potential to the cathodic or anodic direction the external polarization curve can be determined. The internal polarization curve cannot be measured directly by the electrochemical technique since it is impossible to pick up the current separately from the anode and cathode, which exist in an electrode. By analyzing the metallic ions dissolved and the oxidizer reaction the internal polarization curves can be determined.

Anodic or cathodic overpotential is represented by the difference in potential between E_{corr} and E_a or between E_{corr} and E_c and is expressed as n_a or n_c, where

$$n_a = E_{corr} - E_a \qquad (n_a > 0) \tag{5.6}$$

$$n_c = E_{corr} - E_c \qquad (n_c < 0) \tag{5.7}$$

The anodic or cathodic reaction resistance is represented by n_a/i_{corr} or n_c/i_{corr}.

As soon as the cell circuit is formed the corrosion reaction starts:

$$E_c - E_a = |n_c| + i_{corr}R \tag{5.8}$$

where R is the resistance of electrolyte between the anode and cathode. As the current passes through three processes—the anodic process, the cathodic process, and the transit process in the electrolytes—the electromotive force of a corrosion cell is dissipated.

When the electrode is polarized the overpotential n is composed of the activation overpotential n^a and the concentration overpotential n^c:

$$n = n^a + n^c \tag{5.9}$$

The activation overpotential n^a results from the potential energy barrier to be overcome for a charge to cross the electrical double layer at the interface ($M \rightleftharpoons M^{n+} + ne$) and is given as follows:

In the anodic reaction

$$n_a^a = \beta_a \log \frac{i_a}{i_{0a}} \qquad \text{(Tafel equation)} \tag{5.10}$$

$$\beta_a = 2.3\frac{RT}{\alpha nF} \tag{5.11}$$

In the cathodic reaction

$$n_c^a = \beta_c \log \frac{i_c}{i_{0c}} \qquad \text{(Tafel equation)} \tag{5.12}$$

$$\beta_c = -\frac{2.3RT}{(1-\alpha)nF} \tag{5.13}$$

where

n_a^a = activation overpotential in the anodic reaction
n_c^a = activation overpotential in the cathodic reaction
β_a = anodic Tafel coefficient
β_c = cathodic Tafel coefficient
α = transfer coefficient
i_a = anodic current density
i_c = cathodic current density
i_{0a} = exchange current density of anode
i_{0c} = exchange current density of cathode

The degree of contribution of electrical energy for the activation energy in the electrode reaction ($0 < \alpha < 1$) is indicated by the energy transfer factor, α, which in most cases is in the range of 0.3–0.7. The exchange current density, i_{0a} or i_{0c}, is the flux of charge that passes through the electrical double layer at the single-equilibrium potential, E_a or E_c. A linear relationship exists between n^a and log i_a or log i_c. The Tafel coefficient β_a or β_c is the slope $dn^a/d(\log i_a$ or $\log i_c)$ of the polarization curve. Therefore, β is one of the important factors controlling the corrosion rate.

The electrode reaction at the low reaction rate is controlled by the activation overpotential. One of the processes controlled by the activation overpotential is the cathodic reaction (in the acid solution):

$$2H + 2e \rightarrow H_2 \tag{5.14}$$

Table 5.2 shows the hydrogen overpotentials of various metals. The activation overpotential varies with the kind of metal and the electrolytic condition. Metal dissolution and metal ion deposition are usually controlled by the activation overpotential.

The anodic overpotential, n_a, is given by:

$$n_a = \beta_a \log \frac{i_a}{i_{0a}} \tag{5.15}$$

At high reaction rates the concentration overpotential, n^c, becomes the controlling factor in the electrode reaction. In this case, the electrode reaction is controlled by mass transfer process, which is the diffusion rates of reactive species. The diffusion current, i, is given as follows:

$$i = \frac{nFD(C - C_0)}{\delta} \tag{5.16}$$

Table 5.2 Hydrogen Overpotentials of Various Metals

Metal	Temp. (°C)	Solutions	Hydrogen overpotential $\lvert n_0 \rvert$ (V/mA/cm^2)	Tafel coefficient $\lvert \beta_c \rvert$ (V)	Exchange current density $\lvert i_{0c} \rvert$ (A/cm^2)
Pt smooth	20	1 N HCl	0.00	0.03	10^{-3}
Mo	20	1 N HCl	0.12	0.04	10^{-6}
Au	20	1 N HCl	0.15	0.05	10^{-6}
Ag	20	0.1 N HCl	0.30	0.09	5×10^{-7}
Ni	20	0.1 N HCl	0.31	0.10	8×10^{-7}
Bi	20	1 N CCl	0.40	0.10	10^{-7}
Fe	16	1 N HCl	0.45	0.15	10^{-6}
Cu	20	0.1 N HCl	0.44	0.12	2×10^{-7}
Al	20	2 N H_2SO_4	0.70	0.10	10^{-10}
Sn	20	1 N HCl	0.75	0.15	10^{-8}
Cd	16	1 N HCl	0.80	0.20	10^{-7}
Zn	20	1 N H_2SO_4	0.94	0.12	1.6×10^{-11}
Pb	20	0.01–8 N HCl	1.16	0.12	2×10^{-13}

Source: Ref. 1.

where

i = current density
D = diffusion coefficient
C = concentration of reaction species in the bulk solution
C_0 = concentration of reactive species at the interface
δ = thickness of diffusion layer

When the concentration of the reactive species at the interface is zero, $C_0 = 0$, and the current density reaches a critical value, i_L, called limiting current density:

$$i_L = \frac{nFDC}{\delta} \tag{5.17}$$

from Eqs. (5.16) and (5.17):

$$\frac{C_0}{C} = 1 - \frac{i}{i_L} \tag{5.18}$$

The concentration overpotential is given as follows:

$$n^c = \left[2.3 \frac{RT}{nF}\right] \log \left[\frac{C_0}{C}\right] \tag{5.19}$$

From Eqs. (5.18) and (5.19):

$$N^c = \left[\frac{2.3RT}{nF}\right] \log \left[1 - \frac{i}{i_L}\right] \tag{5.20}$$

As seen in Eq. (5.20), the concentration overpotential increases rapidly as i approaches i_L.

The cathodic reaction is controlled by the activation overpotential, n_c^a, and the concentration overpotential, n_c^c.

The cathodic overpotential is:

$$n_c = n_c^a + n_c^c \tag{5.21}$$

The cathodic overpotential, n_c, can be written in the general form:

$$n_c = n_c \log \frac{i_c}{i_{0c}} + \frac{2.3RT}{nF} \log \left[1 - \frac{i_c}{i_{cL}}\right] \tag{5.22}$$

[see Eqs. (5.13), (5.20), and (5.21)]. In most cases the corrosion rate can be determined from the anodic and cathodic overpotentials since the rate-determining process is determined by the slopes of two polarization curves.

As mentioned previously, the role of the coating is to isolate the substrate from the atmosphere. The coating accomplishes this based on two characteristics of the coating material: (1) the corrosion resistance or stability of the coating material when the coating is formed by the defect-free continuous layer, and (2) the electrochemical action of the coating material when the coating layer has some defect, such as a pore or crack. The mechanism of the corrosion cell can explain the action required of a coating layer. For better understanding, Eq. (5.8) is rewritten as:

$$i_{corr} = \frac{(E_c - E_a) - |n_c| n_a}{R} \tag{5.23}$$

A corrosion-resistant coating is achieved by one of five different methods to decrease i_{corr} based on Eq. (5.23):

1. EMF control protection: decrease in electromotive force $(E_c - E_a)$
2. Cathodic control protection: increase in cathodic overpotential $|n_c|$
3. Anodic control protection: increase in anodic overpotential $|n_a|$
4. Mixed control protection: increase in both anodic and cathodic overpotential, $|n_a|$ and $|n_c|$
5. Resistance control protection: increase in resistance of corrosion cell, R

II. EMF CONTROL PROTECTION

The difference in potential between the cathode and the anode $(E_c - E_a)$ is the EMF of the corrosion cell. It is also the degree of thermodynamic instability of the surface metal for the atmosphere. In other words, the less the EMF the lower the corrosion rate. By covering the surface of the active metal with a continuous layer of a more stable metal, the active metal surface becomes more thermodynamically stable. Dissolved oxygen and hydrogen ions, which are the reactants in the cathodic reaction, are normal oxidizers found in natural environments. In the natural atmosphere the single potential of dissolved oxygen is nearly constant. Because of this, metals with more noble electrode potentials are used as coating materials. These include copper, silver, platinum, gold, and their alloys. A copper coating system provides excellent corrosion resistance under the condition that the defect-free continuous layer covers the surface of the iron substrate. In so doing the EMF of the iron surface is decreased by the copper coating. The corrosion potential is changed from E_{corr} of uncoated iron to E_{corr} of copper by coating with copper. Under this condition the iron corrodes at the low rate of i_{corr} of copper. However, if the iron substrate is exposed to the atmosphere as the result of mechanical damage, the substrate is corroded predominantly at the rate of i_{corr} of exposed copper by coupling iron and copper (galvanic corrosion).

Copper-coated steel is suitable for use in rural areas and urban or industrial areas with minimum pollution by SO_x. Acid conditions produce a soluble corrosion product that results in eventual dissolution of the coatings. Such a condition can exist on roofs in areas adjacent to chimneys that emit acid smoke.

EMF can also be decreased by use of organic coatings, glass lining, porcelain enameling, and temporary coatings with greases and oils.

III. CATHODIC CONTROL PROTECTION

Cathodic control protection protects the substrate by coating with a less noble metal, for which the slopes of cathodic polarization curves are steep. The cathodic overpotential of the surface is increased by coating; therefore, the corrosion potential becomes more negative than that of the substrate. Coating metals used for this purpose are zinc, aluminum, manganese, cadmium, and their alloys. The electrode potentials of these metals are more negative than those of iron and steel. When exposed to the atmosphere these coatings act as sacrificial anodes for the iron and steel substrates.

The protective abilities of coatings are as follows:

1. Original barrier action of coating layer
2. Secondary barrier action of corrosion product layer
3. Galvanic action of coating layer as sacrificial anode

Barrier coatings 1 and 2 predominate as the protective ability even though a sacrificial metal coating is characterized by galvanic action.

Initially the substrate is protected against corrosion by the barrier action of the coating, followed by the barrier action of the corrosion product layer.

Upon exposure to air, aluminum forms a chemically inert Al_2O_3 oxide film that is a rapidly forming self-healing film. Therefore the passive film on aluminum, as well as the corrosion product layer, is a main barrier and leads to a resistant material in natural environments.

On the other hand, the surface oxide film that forms on zinc is not as an effective barrier as the oxide film on aluminum.

Upon exposure in the natural atmosphere many corrosion cells are formed on the surface of the sacrificial metal coating, thereby accelerating corrosion rate. During this period corrosion products are gradually formed and converted to a stable layer. This period may last for several months after which the corrosion rate becomes constant. These corrosion products form the secondary barrier and are amorphous Al_2O_3 on aluminum, and $Zn(OH)_2$ and basic zinc salts on zinc. ZnO, being electrically conductive, loosens the corrosion product layer and therefore does not contribute to formation of the barrier. When pollutants such as CO_2, NaCl, and SO_x are present, basic zinc salts are formed, such as $2ZnCO_3 \cdot 3Zn(OH)_2$ in mild atmospheres, $ZnCl_2 \cdot 6Zn(OH)_2$ in chloride atmospheres, and $ZnSO_4 \cdot 4Zn(OH)_2$ in SO_x atmospheres. How stable each basic zinc salt will be is dependent on the pH and anion concentration of the electrolyte on the zinc. Zinc carbonate forms an effective barrier on steel in mild atmospheres while basic zinc sulfate and chloride dissolve with decreasing pH of the electrolyte. The basic zinc sulfate is restricted under atmospheric conditions in its efforts to act as a barrier since the pH value of rain in an SO_x atmosphere is usually low, in the area of less than 5. In a chloride environment the value of pH in the electrolyte is not as low as in the SO_x atmosphere; therefore, a secondary barrier will form. However, in a severe chloride environment the zinc-coating layer will corrode in spite of the existence of basic zinc chloride on the surface.

A. Galvanic Action of Coating Layer

Sacrificial metal coatings protect the substrate metal by means of galvanic action. When the base metal is exposed to the atmosphere as a result of mechanical damage or the like, the exposed small portion of base metal is polarized cathodically to the corrosion potential of the coating layer. As a result little corrosion takes place on the exposed metal. A galvanic couple is formed between the exposed part of the base metal and the surrounding coating metal. Sacrificial metals are more negative in electrochemical potential than other metals such as iron or steel. Therefore, the sacrificial metal coating acts as an anode and the exposed base metal acts as a cathode. The corrosion potentials of sacrificial metals and steel in 3% sodium chloride solution are shown in Table 5.3. The exposed metal is protected against corrosion while the dissolution of the coating layer around the defect is accelerated. This type of reaction of sacrificial metal coatings is known as galvanic or cathodic protection. In addition, the defects are protected by a secondary barrier of corrosion products of the coating layer. (For the schematic illustration of the galvanic action of a sacrificial metal coating, see Figure 5.2.)

Compared with galvanized steel, aluminized steel provides better corrosion resistance, but in the atmosphere, such as in rural or industrial areas, the aluminum coating does not act as a sacrificial anode. Rust forms at a

Table 5.3 Corrosion Potentials of Sacrificial Metals in a 3% NaCl Solution

Metal	Corrosion potential (V, SCE)
Mn	−1.50
Zn	−1.03
Al	−0.79
Cd	−0.70
Steel	−0.61

Source: Ref. 1.

defect in the coating, such as scratches during the initial time of exposure, but the area of red dust does not expand.

B. Corrosion Loss of Metal Coatings

The loss of metal coating resulting from corrosion determines the service life of the coating. The degree of loss is dependent on the time of wetness on the metal surface and the type of concentration of pollutants in the atmosphere. Table 5.4 shows the average corrosion losses of zinc, aluminum, and 55% Al-Zn coatings in various locations and atmospheres. The losses were calculated from the mean values of time of wetness and the average corrosion rate during wet duration. The time of wetness of walls is 40% of that of roofs. Coating metals and coating thicknesses can be decided from Table 5.4 since the corrosion losses of zinc, aluminum, and Al-Zn alloy are proportional to exposure time.

As can be seen from the table, a G90 sheet, which has a 1-mil zinc coating, cannot be used for a roof having a durability of 10 years in any atmosphere except in a rural area. Were this sheet to be used in an urban, marine, or industrial atmosphere, it would have to be painted for protection.

Aluminum and 55% Al-Zn alloy provide galvanic protection for the steel substrate. In rural and industrial atmospheres, an aluminum coating does not act as a sacrificial anode. However, in a chloride atmosphere, such as a marine area, it does act as a sacrificial anode.

C. Application

The choice as to which sacrificial metal coating to use will be based on the environment to which it will be exposed and the service life required. The

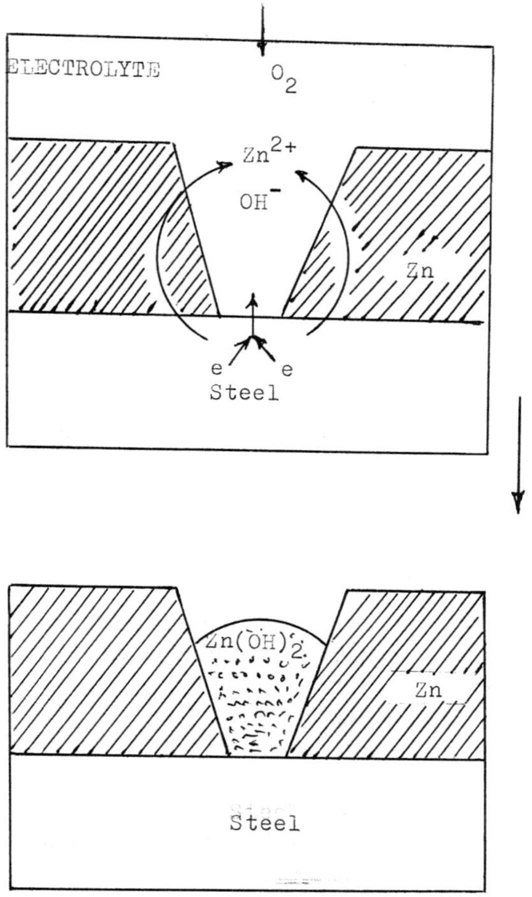

Figure 5.2 Schematic illustration of galvanic action sacrificial metal coating. (From Ref. 1.)

service life required will also determine the coating thickness to be applied, which in turn will influence the coating process to be used. Sacrificial metal coatings have been used successfully for roofs, walls, ducts, shutters, doors, and window frames in the housing industry; on structural materials such as transmission towers, structural member of a bridge, antenna, chimney structure, grandstand, steel frames, high-strength steel bolts, guard rails, corrugated steel pipe, stadium seats, bridge I beams, footway bridges, road bridges, and fencing.

Table 5.4 Average Corrosion Losses of Sacrificial Metal Coatings for 10 Years

Location	Atmosphere	Average corrosion loss (mil[a]/10 yr)					
		Zinc		55% Al-Zn		Aluminum	
		Roof	Wall	Roof	Wall	Roof	Wall
Inland	Rural	0.42	0.17	0.15	0.06	0.06	0.02
	Urban	1.48	0.59				
	Industrial	1.40	0.56	0.15	0.06	0.06	0.02
	Severe industrial	1.59	0.64				
Inland, shore of lake or marsh	Rural	0.59	0.24	0.20	0.08	0.07	0.03
	Urban	1.97	0.79				
	Industrial	1.40	0.56	0.20	0.08	0.08	0.03
	Severe industrial	2.12	0.85				
Coast	Rural	0.74	0.23	0.25	0.10	0.08	0.04
	Urban	2.47	0.99				
	Industrial	1.75	0.70	0.25	0.10	0.10	0.04
	Severe industrial	2.65	1.06				
Seashore	Severe industrial	2.06	0.82	0.46	0.18	0.19	0.07

[a]1 mil = 25.4 μm.
Source: Ref. 1.

IV. ANODIC CONTROL PROTECTION

Noble metal coatings provide anodic control protection. They are usually used where corrosion protection and decorative appearance are required. Nickel, chromium, tin, lead, and their alloys are the coating metals that provide anodic control protection.

A. Mechanism of Corrosion Protection

Noble metal coatings can be either of a single-layer or a multilayer type. Although the noble metal is itself corrosion-resistant, a coating of a noble metal contains such defects as pores and cracks. To overcome these defects multilayer coatings were developed.

B. Single-Layer Coatings

Single-layer noble metal coatings provide corrosion protection as a result of the original barrier action of the noble metal. With the exception of lead, a secondary barrier of corrosion products is not formed. Noble metals do not provide cathodic protection to steel substrates in natural atmospheres because the corrosion potential of the noble metal is more noble than that of steel (see Table 5.5). The service life of a single-layer coating is affected by any discontinuity in the coating, such as that caused by pores and cracks. For nickel to form a protective barrier the coating thickness must be greater than 30 μm to ensure absence of defects. The surface of a bright nickel coating will remain bright in a clean atmosphere but will change to a dull color when exposed in an SO_x atmosphere.

Table 5.5 Corrosion Potentials of Noble Metals

	Corrosion potential (V, SCE)	
pH	2.9	6.5
Chromium	−0.119	−0.186
Nickel	−0.412	−0.430
Tin	−0.486	−0.554
Lead	−0.435	−0.637
Steel	−0.636	−0.505

Source: Ref. 1.

Chromium coatings are applied as a thin layer to maintain bright, tarnish-free surface. Cracking of chromium coatings begins at a thickness of 0.5 μm after which a network of fine cracks is formed.

For protection of steel in an SO_x atmosphere lead and its alloys (5–10% tin) coatings are employed. Pitting will occur in the lead coating at the time of initial exposure, but the pits are self-healed and then the lead surface is protected by the formation of insoluble lead sulfate.

Tin coating of metal substrates is not normally used for corrosion protection in natural atmospheres because the coating tends to darken and stain when exposed.

C. Multilayer Coatings

There are three types of nickel coatings—bright, semibright, and dull-bright. The differences between the coatings is in the quantity of sulfur contained in them as shown below:

Bright nickel deposits > 0.04% sulfur
Semibright nickel deposits < 0.005% sulfur
Dull nickel deposits < 0.001% sulfur

The corrosion potentials of the nickel deposits are dependent on the sulfur content. Figure 5.3 shows the effect of sulfur content on the corrosion potential of a nickel deposit.

As the sulfur content increases, the corrosion potential of a nickel deposit becomes more negative. A bright nickel coating is less protective than a semibright or dull nickel layer coating. The difference in the potential of bright nickel and semibright nickel deposits is more than 50 mV.

Use is made of the differences in the potential in the application of multilayer coatings. The more negative bright nickel deposits are used as sacrificial intermediate layers. When bright nickel is used as an intermediate layer the corrosion behavior is characterized by a sideways diversion. Pitting corrosion is diverted laterally when it reaches the more noble semibright nickel deposit. Thus the corrosion behavior of bright nickel prolongs the time for pitting penetration to reach the base metal.

The most negative of all nickel deposits is trinickel. In the triplex layer coating system a coating of trinickel approximately 1 μm thick, containing 0.1–0.25% sulfur, is applied between bright nickel and semibright nickel deposits. The high-sulfur nickel layer dissolves preferentially, even when pitting corrosion reaches the surface of the semibright nickel deposit. Since the high-sulfur layer reacts with the bright nickel layer, pitting corrosion does not penetrate the high-sulfur nickel layer in the tunneling form. The

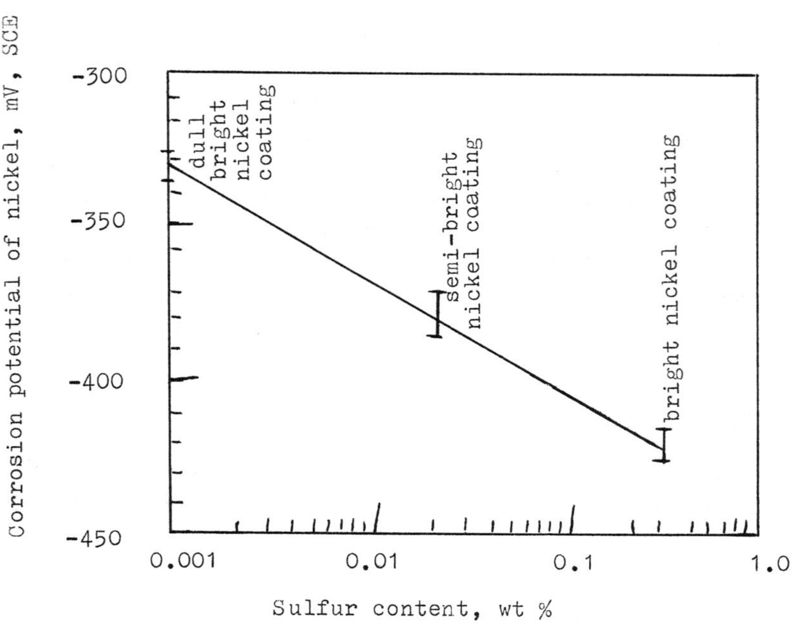

Figure 5.3 Effect of sulfur content on the corrosion potential of nickel deposit. (From Ref. 1.)

application of a high-sulfur nickel strike definitely improves the protective ability of a multilayer nickel coating.

D. Atmospheric Protection

Single-layer coatings of nickel and chromium provide less corrosion resistance than the pure metal. This is due to discontinuities in the coating. The electroless plating process of applying a single-layer coating provides a more corrosion-resistant coating than does the process of electroplating. This is due to the formation of fewer discontinuous deposits. In the production of multilayered coatings the roles of the various deposits are as follows:

1. Microdiscontinuous chromium deposit prevents the concentration of pitting corrosion.
2. Bright nickel deposits with trinickel strike acts as a sacrificial anode.
3. Semibright nickel strike deposit forms a barrier against pitting corrosion penetrating to the substrate.

Duplex coatings are superior to bright nickel coatings in many atmospheres. Since the chromium deposit is the topmost layer in a multilayer coating, the durability of the chromium deposit controls the corrosion behavior of the duplex nickel during the initial stage of exposure. Duplex nickel coatings provide excellent corrosion resistance in natural atmospheres.

As the nickel layer thickness is increased in a multilayer nickel coating the corrosion resistance of the coating is improved.

V. EMF CONTROL PROTECTION

Copper, silver, platinum, gold, and their alloys are the coating metals belonging to this category. The standard single potentials of these metals are more noble than those of hydrogen. Since most of these are precious metals, they are used primarily for decorative appearance and electrical conduction. The one exception is copper.

A. Mechanism of Corrosion Protection

The corrosion protection provided by a copper coating is twofold, consisting of an original barrier action of the coating layer and a secondary barrier action of corrosion products. The low EMF of copper is responsible for the formation of the original barrier action. The electrochemical reactions in the corrosion cells on copper are as follows:

Anodic reaction:

$$Cu \rightarrow Cu^{+} + e$$

$$Cu \rightarrow Cu^{2+} + 2e$$

Cathodic reaction:

$$O_2 + 2H_2O + 4e \rightarrow 4OH^-$$

Chloride ions in a natural environment stabilize cuprous ions. Cupric ions are more stable. Since the EMF of corrosion on copper is less than that on iron, the reactivity of a steel surface is decreased by coating it with copper.

Over a period of time corrosion products gradually build up a secondary layer against corrosion. Initially a cuprous oxide layer is formed, followed by the copper surface covered with basic copper salts. Pollutants in the atmosphere determine the formation of basic copper salts as follows:

Mild atmosphere	Malachite $CuCO_3{:}Cu(OH)_2$
SO_x atmosphere	Brochanite $CuSO_4{:}3Cu(OH)_2$
Chloride atmosphere	Atacamite $CuCl_2{:}3Cu(OH)_2$

In most coastal areas the amount of sulfates in the atmosphere exceed the amount of chlorides. As a layer of copper salts grows on the surface of the corrosion product layer the protective ability of the corrosion layer is increased. As the exposure time increases the average corrosion rate of copper gradually decreases. After 20 years the corrosion rate of copper is reduced to half the value of the first year as a result of the secondary barrier of corrosion products.

B. Atmospheric Protection

The initial corrosion rate of a copper coating is dependent on atmospheric conditions such as time of wetness and the types and amount of pollutants. Time of wetness is the most important factor affecting the corrosion rate of copper. The corrosion rate of copper usually obeys parabolic law:

$$M^2 = k\mathrm{t}$$

where

M = mass increase
k = a constant
t = exposure time

Accordingly, the average corrosion rate decreases with increased exposure time, which means that the surface of the copper is covered with basic salts by degrees and thereafter the corrosion rate approaches a constant value.

Copson (2) conducted 20-year exposure tests and found the average corrosion rate of copper to be as follows:

0.0034 mil/year in dry rural atmospheres
0.0143 mil/year in rural atmospheres
0.0476–0.0515 mil/year in industrial atmospheres
0.0198–0.0562 mil/year in marine atmospheres

Until the base metal is exposed the corrosion process of a copper-coated layer is similar to that of copper plate. Galvanic corrosion of copper-coated steel is induced when the steel substrate is exposed. However, in the case of copper-coated stainless steel the occurrence of galvanic action is dependent on the composition of the stainless steel.

In a chloride atmosphere galvanic tunneling at cut edges and galvanic pitting takes place at the pores in copper layers on types 409 and 430 stainless steels, whereas in SO_x atmospheres uniform corrosion takes place on the copper coating.

Copper coatings are used both for decorative purposes and for corrosion protection from the atmosphere. Copper-coated steels are used as roofs, flashings, leaders, gutters, and architectural trim. Copper undercoats also improve the corrosion resistance of multilayered coatings, specifically in the plating of nickel and chromium.

REFERENCES

1. I Suzuki. Corrosion Resistant Coatings Technology. New York: Marcel Dekker, 1989.
2. HR Copson. Report of Subcommittee VI of Committee B-3 on atmospheric corrosion of nonferrous metals, ASTM annual meeting, Atlantic City, June 25, 1955.

6 Atmospheric Corrosion of Conversion Coatings

Conversion coatings are coatings either formed on metal naturally by reaction with the environment or man-made using chemical or electrochemical treatment. In a conversion coating the substrate metal provides ions that become part of the protective coating. A conversion coating in itself does not necessarily provide sufficient corrosion protection. Conversion coatings on steel such as phosphate coatings, chromate coatings, and mixed oxide coatings are generally used beneath a paint or organic coating, whereas anodized coatings on aluminum are generally used with no further protection.

Pretreatment layers (conversion coatings) are used for a variety of reasons:

1. To obtain electrically insulating barrier layers
2. To improve the adherence of the organic layer
3. To provide a uniform grease-free surface
4. To provide active corrosion inhibition by passivating the metallic substrate or by reducing the rate of the oxygen–reduction reaction

I. OXIDE COATINGS ON STEEL

Oxide coatings on steel can be produced by immersion in hot concentrated alkali solutions containing persulfates, nitrates, or chlorates. They can also be prepared by controlled high-temperature oxidation in air. These coatings consist mostly of Fe_3O_4 and are blue, brown, or black in color, and in themselves offer no protection against corrosion. Some protection can be obtained by rubbing with inhibiting oils or waxes as is frequently done with gun barrels.

II. OXIDE COATINGS ON ALUMINUM

Oxide coatings on aluminum are formed by anodic oxidation of aluminum in a suitable electrolyte. The process is known as anodizing and the finished product is known as anodized aluminum.

Anodizing is the electrochemical treatment of a metal in an electrolyte where the metal serves as an anode. Many electrolytes can be used but the four main solutions used are sulfuric acid, chromic acid, boric acid, and oxalic acid. Of these four, sulfuric acid is most widely used.

Porous or nonporous films can be formed but the porous film is the one most widely used for corrosion protection. Porous oxide films are composed of two layers: an outer porous layer of duplex layer structure, and an inner nonporous layer (barrier layer). The main compounds of porous oxide films are amorphous Al_2O_3, γ-Al_2O_3, γ-Al_2O_3, α-Al_2O_3, and other compounds depending on the electrolyte used.

The porous structure has a very strong adsorbing ability. This permits the surface of an anodic film to be dyed, but it can also be contaminated. To prevent the formation of corrosion cells, the surface must be sealed. This is accomplished using hot water or steam, which seals the pores by the formation of boehmite ($Al_2O_3 \cdot H_2O$) or bayerite ($Al_2O_3 \cdot H_2O$).

The anodic coating is devoid of coloring and is clear and transparent on aluminum. Alloys containing manganese or silicon and the heterogeneous aluminum-magnesium alloys yield coatings that range from gray to brown and may be patchy in some places. Coatings formed in sulfuric acid are especially suitable for the protection of electropolished surfaces.

Compared to the coatings prepared in sulfuric acid, those prepared in chromic acid are unattractive. They are generally iridescent, opaque, and gray. The quality is dependent on the purity and concentration of the electrolyte. With a sulfate concentration of 0.03% in the electrolyte a colorless and transparent coating will result. These coatings are difficult to dye since they are generally thin and of low porosity. At elevated temperatures and in concentrated solutions black coatings can be obtained. If titanium, zirconium, and thallium compounds are added to the electrolyte an attractive opaque surface will result.

Chromic acid anodizing is the only anodizing process that can be safely employed on structures containing blind holes, crevices, or difficult-to-rinse areas. This process is used extensively in aircraft finishing. While sulfuric acid electrolytes may produce decreases in fatigue strength, chromic acid anodizing generally increases the fatigue strength.

Boric acid electrolytes produce a coating that is essentially nonporous. The film is iridescent and oxides in the range of 2500–7500 Å are generally obtained.

Protective and decorative coatings are also obtained from oxalic and other organic acid electrolytes. The unsealed coatings are usually yellow. They are harder and more abrasion-resistant than conventional sulfuric acid coatings, but they are not superior to the specially hard coatings obtained under special conditions in solutions based on sulfuric acid.

The anodic films formed in sulfuric acid contain considerable amounts of sulfate ion and small amounts of water, whereas those formed in chromic acid and boric acid electrolytes are nearly pure Al_2O_3. While 15% sulfate is common, the actual sulfate concentration varies considerably depending on the conditions of formation. At higher formation temperatures and lower current densities the sulfate concentration is less. The sulfate ion is tightly bound since only a small fraction is removed by extensive leaching. Anodic films formed in oxalic acid and phosphoric acid also contain included anions, but the concentration is much lower than that in films formed in sulfuric acid. During the sealing in boiling water, very little of the anion is removed.

Anodized films consist of two major components, the nonporous barrier layer adjoining the metal and a porous layer extending from the barrier layer to the outer surface of the films. Both barrier and porous layers are obtained in sulfuric, chromic, and oxalic acid electrolytes, whereas boric acid electrolytes yield only barrier type films.

A. Protective Ability

The service life of anodic films are determined by the environmental conditions to which they are exposed, primarily SO_x gases and depositions.

Anodized aluminum is subject to pitting and galvanic corrosion. Contact with other metals such as copper, iron, and steel will cause galvanic corrosion. When anodized aluminum is in contact with mortar or concrete, corrosion is induced since Cl^- ions leach from these materials. As discussed previously, atmospheric corrosion of metals is influenced by pollutants and time of wetness. Depositions and SO_x gas are the most aggressive factors promoting atmospheric corrosion of anodized aluminum. Deterioration of anodic films are promoted by deposits of grime, sulfates, and chlorides, since they tend to absorb aggressive gases and moisture, which increases the time of wetness and decreases the pH of the electrolyte at the interface between the deposits and the surface.

Rainfall, while increasing the time of wetness, also has the advantageous effect of cleaning the surface rather than making the surface corrosive. In other words, some deposits are removed by the rain, so that cleaning anodized aluminum surfaces with water is one of the processes to protect these surfaces against corrosion. In order to remove deposits in industrial atmospheres, cleaning must be done using detergents, since these deposits tend to be greasy. In contrast, deposits in marine atmospheres can be removed with clean water because the depositions are mostly soluble chlorides.

Since the most aggressive pollutant for anodic films is SO_x gas, the corrosion behavior of anodized aluminum depends greatly on the SO_x concentration in the atmosphere. As the SO_x concentration increases, the corrosion area increases linearly.

The thickness of the anodic film is an important factor in its effectiveness. The logarithm of an acceptable life in years is approximately proportional to the anodic film thickness, and a minimum anodic film thickness depending on the exposure conditions:

Severe industrial atmosphere	35 μm
Moderately industrial atmosphere	16 μm
Severe marine atmosphere	12 μm

to provide a service life in excess of 10 years.

The corrosion resistance of anodized aluminum alloys is influenced by the environment in the same way as anodized aluminum and varies with the type of alloy. Aluminum alloys that are anodized include 10% Cu-Al, 10% Mg-Al, 13% Si-Al, 5% Fe-Al, 2% Cr-Al, 1% Ti-Al, 4% Fe–8% SiAl, 6.4% Mg–3.7% Si-Al, 9% Zn–3% Mg-Al, and others. When exposed to moderately industrial atmospheres the order of pitting resistance of anodized aluminum alloys is 5052 > 6063 > 1100 > 3003 > 4043 > 6351 > 2014.

Anodic films are colored by the action of its alloy elements, by the electrolyte, or by dyeing to provide a decorative appearance for architectural materials such as building curtain walls and for window frames. For example, the anodic film formed on Al-Si 4043 alloys with a sulfuric acid bath exhibits a gray color, while a gold color is formed in an oxalic acid bath. The oxalic acid film exhibits superior corrosion resistance over the sulfuric acid film.

Weathering causes changes in colored anodic films. Gold and bronze anodic films tend to fade after 2 years of exposure, but the equivalent degree of color change in most anodic films colored by dye takes 5 years of exposure. Anodic films also provide a better base for paint adhesion than bare aluminum. (Refer to Chapter 5 for additional information on anodized aluminum.)

III. CHROMATE COATINGS ON ANODIZED ALUMINUM

Chromate coatings are applied to aluminum, cadmium, copper, iron, magnesium, silver, zinc, and their alloys. The corrosion film is the result of the following three properties:

1. Hexavalent chromium is reduced to form trivalent chromium, which has poor solubility in aqueous media and forms a barrier layer.
2. Hexavalent chromium included in the coating is reduced to trivalent chromium to repassivate the surface when it is damaged, thereby preventing the formation of hydrogen gas.
3. The rate of the cathodic oxygen reaction is strongly reduced.

Chromic acid, being a strong oxidizing agent, passivates a metal surface, so that a chromate film will not develop. To overcome this, activators such as sulfate, formates, and chlorides are added to the chromate bath to inhibit the passivation of the base metal. The primary components of chromate baths are chromic acid and hexavalent chromium salts.

When a metal is immersed in a chromate bath containing sulfuric acid ($H_2Cr_2O_7 + H_2SO_4$) a chromate film is formed by the following reactions:

1. Corrosion of metal by sulfuric acid

   Anodic reaction (M = metal):

$$M \rightarrow M^{n+} + ne$$

Cathodic reaction:

$$2H^+ + 2e \rightarrow H_2$$

2. Reduction of hexavalent chromium ($HCr_2O_7^-$) to trivalent chromium

$$2HCr_2O_7^- \rightarrow 2CrO_4^{2-} + 2H^+$$

$$CrO_4^{2-} + 8H^+ + 3e \rightarrow Cr^{3+} + 4H_2O$$

$$3H_2 + HCr_2O_7^- \rightarrow 2Cr(OH_3) + OH^-$$

3. Precipitation of trivalent and hexavalent chromium compounds

$$2Cr(OH)_3 + CrO_4^{2-} + 2H^+ \rightarrow Cr(OH)_3 \cdot Cr(OH)CrO_4$$

$$\rightarrow Cr_2O_3 \cdot CrO_3 \cdot nH_2O$$

The inhibiting effect of the soluble hexavalent compounds in the film provides another protective action. Chromate films commonly exhibit an iridescent color. When a treated metal is exposed in a natural or salt spray atmosphere the color of the surface changes from yellow to green, or becomes clear. This indicates the reduction of soluble hexavalent chromium to insoluble trivalent chromium. When in contact with moisture the hexavalent chromium slowly leaches, providing a self-healing action for defects in the coatings. The leached hexavalent chromium acts as an anodic inhibitor. Dissolved hexavalent chromiums are adsorbed by the defects and form passive films that are composed of mixtures of insoluble trivalent chromium compounds and oxides of the base metal.

There are two types of processes whereby chromate conversion coatings may be produced: chromic acid processes (just discussed) and chromic acid–phosphoric acid processes. The overall reaction for the formation of a phosphoric acid process coating is

$$2H_2Cr_2O_7 + 10H_3PO_4 + 12HF + 4Al \rightarrow CrPO_4$$

$$+ 4AlF_3 + 3Cr(H_2PO_4)_3 + 14H_2O$$

A greenish layer results that consists primarily of hydrated chromium phosphate with hydrated chromium oxide, concentrated toward the metal. Aluminum oxides and other aluminum salts are present at the conversion coating–aluminum substrate interface. Since these coatings do not contain any hexavalent chromium compounds they do not provide self-healing action for defects. The thinner chromic acid–phosphoric acid conversion coatings are an excellent base for paint layers and thicker coatings are often applied unpainted.

A. Protective Ability

Chromate conversion coatings provide good corrosion resistance in mild atmospheres and provide good surface appearance. They are usually applied to zinc and its alloy-coated steel sheets (especially to protect them against staining during storage), and to products of zinc die castings, aluminum and its alloys, and magnesium and its alloys. Chromate coatings are also excellent bases for organic films.

Chromate conversion coatings consist of soluble hexavalent chromium compounds and insoluble trivalent chromium compounds, and in some instances an oxide of the base metal. Most coatings are amorphous gel-like precipitates providing excellent continuity.

When exposed to the atmosphere, hexavalent chromium leaches slowly from the film with the result that the surface coloration changes from iridescent yellow to either a green color or to clear. As the leaching proceeds the structure of the film shows an increase in the insoluble trivalent chromium compounds. Deterioration of the film takes place only after months or even years in a natural atmosphere.

The longer the time of wetness the shorter the life of the coating since chromate coatings absorb moisture, and moisture results in the leaching of hexavalent chromium. However, the base metal is protected against corrosion as long as the leaching continues. Leaching of hexavalent chromium is accelerated in a severe corrosive atmosphere. Because of the leaching the service life of a chromate coating is dependent on the coating thickness.

The leaching rate of the hexavalent chromium varies with the aging process. Aging tends to convert hexavalent chromium compounds to more stable compounds that do not tend to leach. Exposure of chromate coatings to direct sunlight results in cracking, thereby decreasing the life of the coating.

In general, the leaching rate is commonly low in natural environments but the presence of pollutants will increase the deterioration rate of the film, particularly chloride ions.

IV. PHOSPHATE COATINGS

Phosphate coatings are applied to iron and steel, zinc, aluminum, and their alloys. Phosphate coatings on steel are also known as "parkerizing" or "bonderizing." These coatings are secondary barriers that are formed by immersion, spraying, or other techniques with a phosphate solution. The secondary barriers consist of crystalline or amorphous phosphates of the metal to be coated, or of metal ions in the phosphating solution. Zinc, iron, and manganese phosphates are typical coating materials.

Phosphate films are formed in the same manner as chromate coatings—by the dissolution of the base metal and the precipitation of phosphate films. The reactions of iron and steel in a zinc phosphate solution are as follows:

1. Dissolution of base metal:

$$Fe + 2H_3PO_4 \rightarrow Fe(H_2PO_4)_2 + H_2$$

Promotion by the activator:

$$2Fe + 2H_2PO_4^- + 2H^+ + 3NO_2 \rightarrow 2FePO_4 + 3H_2O + 3NO$$

$$4Fe + 3Z_n^{2+} + 6H_2PO^- + 6NO_2 \rightarrow$$

$$4FePO_4 + Zn_3(PO_4)_2 + 6H_2O + 6NO$$

2. Precipitation of phosphate films:

$$2ZnFe(PO_4)_2 + Fe(H_2PO_4)_2 + 4H_2O \rightarrow \underbrace{Zn_2Fe(PO_4)_2 \cdot 4H_2O}_{\text{Phosphophyllite}} + 4H_3PO_4$$

$$3Zn(H_2PO_4)_2 + 4H_2O \rightarrow \underbrace{Zn_3(PO_4)_2 \cdot 4H_2O}_{\text{Hopeite}} + 4H_3PO_4$$

In this case the phosphate film consists of phosphophyllite and hopeite. The solubility and continuity of the conversion coating determines the effectiveness of the barrier action. Phosphate films are nonelectric conductive compounds and are stable in neutral atmospheres, their solubilities being the lowest at pH range 6–8. Since the phosphate films deposit on the cathodic areas and not on anodic sites, their continuity is not as good as those of anodic oxides and chromate films. The anodic sites remain as pinholes.

A. Protective Ability

Phosphate coatings do not offer as much corrosion protection as anodic oxide or chromate coatings because of the small pinholes remaining at the anodic sites. However, this surface condition of the phosphate film, which contains very small defects distributed uniformly, provides excellent adhesion for lacquer films and makes the electrodeposition of primers possible.

The presence of pinholes in phosphate films results in rapid rusting under severe corrosive conditions. Phosphate coatings by themselves are not normally used to provide corrosion protection. Their main advantage is in providing an excellent surface for the bonding of paint films.

Water and oxygen diffuse through paint film at a relatively high rate. The protective ability of a paint-phosphate film is influenced by the properties of the phosphate film because the film is in contact with the diffused water and oxygen. Film properties of concern are solubility, uniformity, and porosity.

The solubility of phosphate films varies with pH, the type of phosphate, and its morphology. In the alkaline range of pH >10 the solubility of phosphate film is directly related to the adhesion of paint film. Formation of corrosion cells under the paint film causes an increase in pH at the cathodic area since the cathodic reactions produce OH^- ions and pH at the interface between the steel and paint film reaches 13.5. During the cathodic electrodeposition process for primers, pH at the surfaces reaches a value of 10–12, whereas in the anodic electrodeposition process pH at the surface is 3–4.

Solubility of a phosphate film is low in the pH range of 6–8 and increases in lower or higher pH ranges. Therefore, the solubility of a phosphate film in the pH range of approximately 10 affects paint adhesion.

Solubility also varies depending on the coating process, whether by spray or immersion. In other words, the morphologies of phosphate films result in different solubilities. The denser the phosphate film, the lower will be its solubility. A looser film formed by spraying has a greater solubility than one formed by immersion, which is denser.

The specific phosphate compound also determines the solubility. For example, phosphated steel provides better adhesion for paint films than does phosphated zinc. Phosphates formed on steel are primarily phosphophyllite and those on zinc are hopeite.

Phosphophyllite has a lower solubility than hopeite in low and high pH ranges. The solubility of hopeite increases sharply around pH 11, whereas the solubility of phosphophyllite remains low.

The uniformity of phosphate films is influenced by the surface condition of the base metal and the nucleation of phosphates. Carbonaceous residues on steel surfaces resulting from rolling lubricants during the annealing of steels have an effect on the corrosion performance of paint-phosphate-coated steels. The porosity of a phosphate film is increased as a result of carbon contamination on the steel surface, which promotes the formation of defects in the film.

V. PHOSPHATE-CHROMATE COATINGS

A chromate film is normally very thin and exhibits poor resistance to abrasion. Phosphate conversion coatings do not provide adequate corrosion protection. The phosphate-chromate coatings were developed to improve the quality of both chromate and phosphate coatings. Combining the two different conversion coatings increases corrosion and abrasion resistance.

Phosphate-chromate coatings are composed of dense amorphous layers and do not contain hexavalent chromium.

REFERENCES

1. HH Uhlig. Corrosion and Corrosion Control. New York: John Wiley and Sons, 1963.

2. JHW deWit. Inorganic and Organic Coatings. In: P Marcus, J Oudar, eds. Corrosion Mechanisms in Theory and Practice. New York: Marcel Dekker, 1995, pp. 581–624.
3. H Leidheiser Jr. Coatings. In: F Mansfield, ed. Corrosion Mechanisms. New York: Marcel Dekker, 1987, pp. 165–209.
4. I Suzuki. Corrosion-Resistant Coatings Technology. New York: Marcel Dekker, 1989.

7
Atmospheric Corrosion of Polymeric Coatings

There are three broad classes of polymeric coatings: lacquers, varnishes, and paints. Varnishes are materials that are solutions of either a resin alone in a solvent (spirit varnishes), or an oil and a resin together in a solvent (oleoresinous varnishes). A lacquer is generally considered to be a material whose basic film former is nitrocellulose, cellulose acetate-butyrate, ethyl cellulose, acrylic resin, or another resin that dries by solvent evaporation. The term *paint* is applied to more complex formulations of a liquid mixture that dries or hardens to form a protective coating.

Corrosion protection provided by polymeric coatings results either from the barrier action of the layer against water, oxygen, and ions, or from active corrosion inhibition provided by pigments in the coating. Coatings that contain large quantities of metallic zinc also provide corrosion protection by galvanic action. The barrier properties of the coating are improved by increased thickness, by the presence of pigments and fillers that increase the diffusion path for water and oxygen, and by the ability to resist degradation. Thicknesses may range from a few tenths of a mil to a quarter of an inch or more, although thicknesses between 2 and 40 mils are the most common (1 mil = 1/1000 in. = 25.4 μm).

Mechanisms by which coatings fail include abrasion and impact; cracking or crazing at low or high temperature; bond breakage within the polymer matrix because of hydrolysis reactions, oxidation, or ultraviolet (UV) light; and freeze–thaw cycling. Such degradation permits corrodents to reach the substrate interface without the necessity for diffusing through the polymer matrix. Under these conditions the otherwise adequate barrier properties of the coating no longer give sufficient protection. To compensate for this, active pigments are often incorporated in the polymer matrix of the primer (the first coating layer near the substrate). These pigments (passivating, blocking, or galvanic) provide protection through an active inhibition mechanism immediately, when water and some corrosive agent reach the metal surface.

I. COMPOSITION OF COATINGS

The composition of the paint (coating) is a major influence on the corrosion protection to be supplied. There are four general classes of ingredients that make up the formulation, i.e., vehicle (binder, resin component), pigment, filler, and additive.

A. Resin Component

The resin component is also known as the vehicle or binder, but resin component is more descriptive. The resin is the film-forming agent of the paint. It forms the matrix of the coating, the continuous polymeric phase in which all other components may be incorporated. Its density and composition largely determine the permeability, corrosion resistance, and UV resistance of the coating.

The most common resins used are vinyls, chlorinated rubber, acrylics, epoxies, urethanes, polyesters, autooxidative crosslinking coatings, and water-soluble resins.

B. Vinyl Resins

Vinyl is the general term denoting any compound containing the vinyl linkage ($-CH=CH_2$) group. However, there are many compounds containing this linkage that are not considered as vinyl coatings, such as styrene and propylene.

Vinyl coatings are primarily considered to be copolymers of vinyl chloride and vinyl acetate having the chemical structure shown in Figure 7.1. A relatively large amount of solvent is required to dissolve a vinyl copolymer high in vinyl chloride content. Because of this, the volume of solids in the solution is relatively low. As a result, most vinyl coatings must be applied in thin (1–1.5 mil) coats. Consequently, a vinyl coating system might require five or more coats. The system is considered highly labor-intensive, even though in the proper environment excellent protection is provided. A formulation has been developed that permits coats of 2–2.5 mil to be applied but this advantage comes at the expense of reduced protection. The thixo-

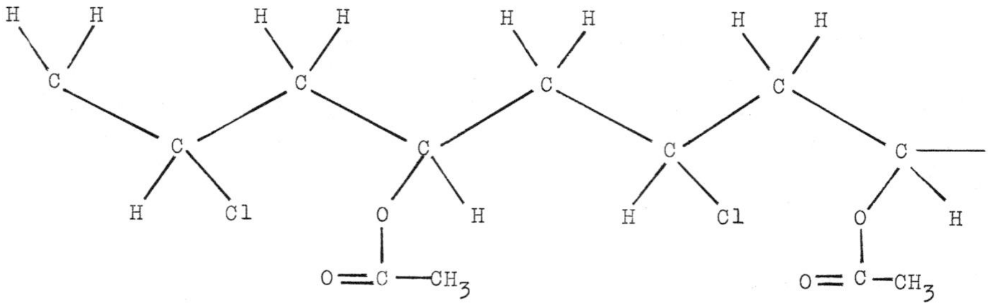

Figure 7.1 Vinyl acetate and vinyl chloride copolymer chemical structure.

tropes, fillers, and additives used to permit the greater thickness are more susceptible to environmental and moisture permeation.

Vinyl coatings find application in industrial atmospheres because of their resistance to acid and alkaline gases and chemicals. They also find application in marine atmospheres because of their resistance to salt sprays.

C. Chlorinated Rubber

Chlorinated rubber resins include those resins produced by the chlorination of both natural and synthetic rubbers. Chlorine is added to unsaturated double bonds until the final product contains approximately 65% chlorine with the chemical structure shown in Figure 7.2. Since the final product is a hard, brittle material with poor adhesion and elasticity, a plastizizer must be added. The volume of solids of the coating is somewhat higher than that of a vinyl; therefore, a suitably protective chlorinated rubber system often consists of only three coats.

Chlorinated rubber paints find application in rural and mountain areas with high humidity and much snow, in urban atmospheres because of their resistance to automobile exhausts and waste gases from factories, and in marine atmospheres because of their resistance to salt spray. The chlorinated rubber paints are degraded by UV light.

D. Acrylics

The acrylics can be formulated as thermoplastic resins, thermosetting resins, and as a water emulsion latex. The resins are formed from polymers of acrylate esters, primarily polymethyl methacrylate and polyethyl acrylate. Since the acrylate resins do not contain tertiary hydrogens attached directly to the polymer backbone chain, they are exceptionally stable to oxygen and

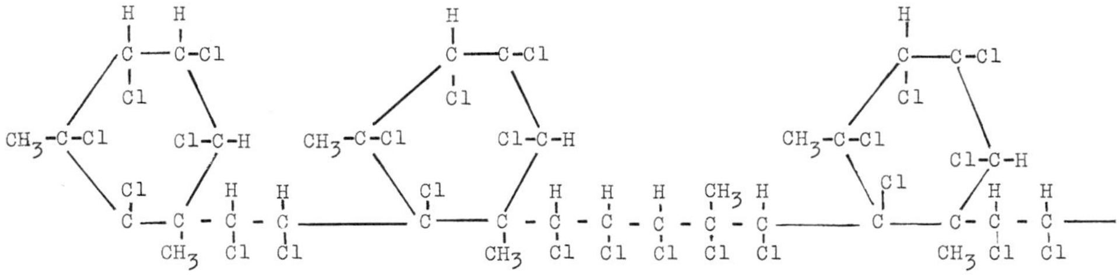

Figure 7.2 Chlorinated rubber chemical structure.

UV light. The repeating units for the methyacrylate and acrylate are as follows:

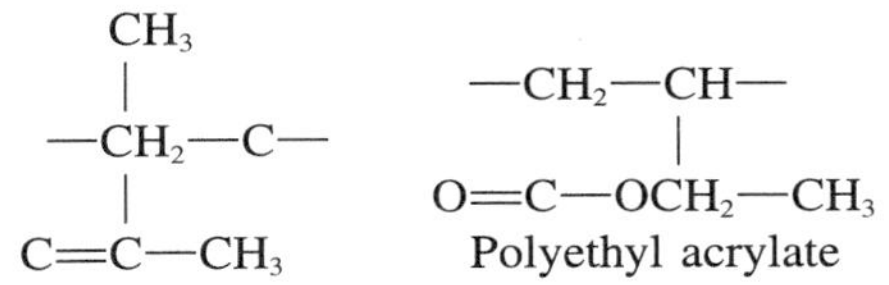

Polymethylate   Polyethyl acrylate

Acrylic resins, particularly the methyacrylates, are somewhat resistant to acids, bases, weak and moderately strong oxidizing agents, and many corrosive industrial gases and fumes.

E. Epoxies

Epoxy resins themselves are not suitable for protective coatings because when pigmented and applied they dry to a hard, brittle film with very poor chemical resistance. When copolymerized with other resins, specifically with those of the amine or polyamide family, or with esterified fatty acids, epoxy resins will form a durable protective coating.

Epoxy coatings, crosslinked with amines or polyamines, are widely used in atmospheric fume environments. Because of the sensitivity of coal tar pitch to UV light, coal tar epoxies are not normally used in atmospheric exposures.

F. Urethanes

Urethanes are reaction products of isocyanates with materials having hydroxyl groups and simply contain a substantial number of urethane groups (−N−C−O−) regardless of what the rest of the molecule may be. The hydroxyl-containing side might consist of a variety of compounds including water (moisture-cured urethanes), epoxies, polyesters, acrylics, and drying oils. Epoxy and polyester polyols, though more expensive than the acrylic polyols, are more chemically and moisture-resistant. However, the acrylic polyol when suitably reacted to form a urethane coating is satisfactory for most weathering atmospheres.

Coatings formulated with aliphatic isocyanates cost approximately twice as much as similar coatings formulated with epoxy and aromatic isocyanates. However, they have superior color and gloss retention and are resistant to deterioration by sunlight. Aliphatic urethanes will continue to maintain a glossy "wet" look years after application, whereas other urethane coatings will chalk or yellow somewhat after exposure to the sun. Urethane coatings are not suitable for prolonged exposure to water or strong chemical environments.

G. Polyesters

In the coating industry, polyesters are characterized by resins based on components that introduce unsaturation (−C=C−) directly into the polymer backbone. Typical is the structure of an isophthalic polyester resin, as shown in Figure 7.3. The most common polyester resins are polymerization products of maleic or isophthalic anhydride or their acids. In producing paint, the polyester resin is dissolved in styrene monomer, together with pigment and small amounts of a reaction inhibitor. A free radical initiator, commonly a peroxide, and additional styrene are packaged in another container. When applied the containers are mixed; sometimes, because of the fast initiating reaction (short pot life), they are mixed in an externally mixing or dual-headed spray gun. After being mixed and applied, a relatively fast reaction takes place, resulting in crosslinking and polymerization of the monomeric styrene with the polyester resin.

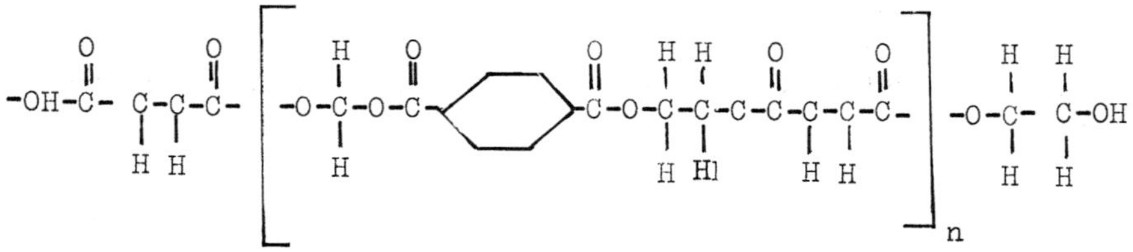

n = 3 to 6

Figure 7.3 Chemical structure of polyester coating.

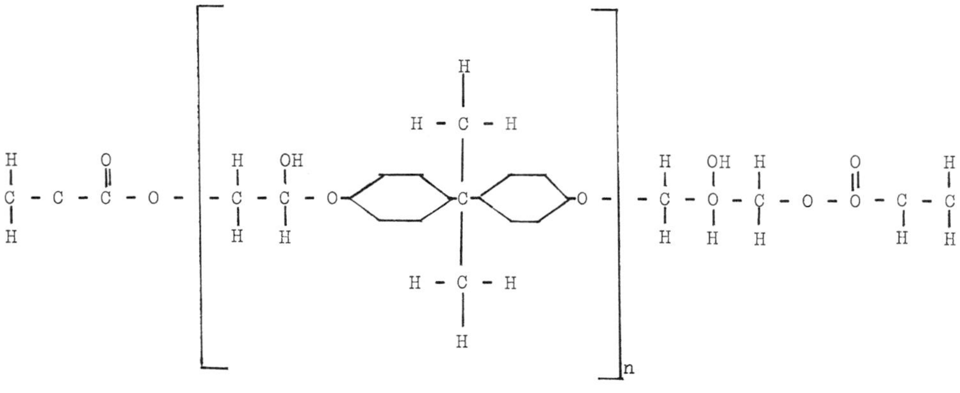

Figure 7.4 Chemical structure of vinyl ester coating.

Another class of coatings, formulated in a similar manner, are the vinyl esters. The vinyl esters derive from a resin based on a reactive end vinyl group that can open and polymerize. The chemical structure is shown in Figure 7.4. The vinyl ester has better moisture and chemical resistance and is more stable than the polyester.

These paints have good resistance to weathering and excellent resistance to acids.

H. Autooxidative Crosslinking Coatings

Autooxidative crosslinking coatings rely on a drying oil with oxygen to introduce crosslinking within the resin and attainment of final film properties. The final coating is formed as a result of the drying oil reacting with the resin, which is combined with pigments and solvents. The paint is packaged in a single can, which can be opened, mixed, and paint applied. Since oxygen reacts with the coating introducing additional crosslinking, final film properties can take weeks or even months to attain.

Autooxidative crosslinking–type paints include alkyds, epoxy esters, oil-modified urethanes, and so on. These are commonly used, when properly formulated, to resist moisture and chemical fume environments. They can be applied over wood, metal, or masonry substrates.

The advantage of these coatings is their ease of application, excellent adhesion, relatively good environmental resistance (in all but immersion and high chemical fume environment), and tolerance for poorer surface preparation. Their major disadvantage, compared with other synthetic resins, is their lessened moisture and chemical resistance.

I. Water-Soluble Resins and Emulsion Coatings

Any type of resin can be made water-soluble by introducing sufficient carboxyl groups into the polymer. These groups are then neutralized with a volatile base such as ammonia or an amine rendering the resin a polymeric salt, soluble in water or water/ether-alcohol mixtures. The major disadvantage to such resins is that polymers designed to be dissolved in water will remain permanently sensitive to water. Because of this they are not widely used industrially.

However, water emulsion coatings are widely accepted. The water-based latex emulsions are formulated with high molecular weight resins in the form of microscopically fine particles of high molecular weight copolymers of polyvinyl chloride, or polyvinyl acetate, styrene-butadiene, acrylic esters or other resins combined with pigments, plasticizers, UV stabilizers, and other ingredients. Water-based epoxy formulations are also available. The epoxy resin and a polyamide copolymer are emulsified and packaged in separate containers. After mixing, coalescence, and ultimate drying, the polyamide reacts with the epoxy to form the final film.

The major advantage of the emulsion coatings is their ease of clean-up. Their disadvantage is the permanent water sensitivity of the coating, making them unsuitable for use in continually wet environments. Water-based coatings cannot be applied to blast-cleaned steel surfaces with the same confidence as solvent-based coatings.

J. Pigments

Pigments serve several functions as part of the coating; they reinforce the film structurally, provide color and opacity, and serve special purposes for metal protection.

Typical materials used as color or hiding pigments as well as to provide aesthetic value, retention of gloss and color, and to help with the film structure and impermeability are iron oxides, titanium dioxide, carbon or lamp black, and others.

Pigments must be somewhat resistant to the environment and must be compatible with the resin. For example, calcium carbonate, which is attacked by acid, should not be used in an acid environment. Water-soluble salts are corrosion promoters, so special low-salt-containing pigments are used in primers for steel.

Primers contain one of three kinds of pigments for special protective properties as follows:

1. *Inert or chemically resistant.* These are used to form barrier coatings in severe environments, such as pH less than 5 or greater than 10.

They are also used as nonreactive extender, hiding, or color pigments in neutral environments.
2. *Active*. Leads, chromates, or other inhibitive pigments are used in linseed oil/alkyd primers.
3. *Galvanically sacrificial*. Zinc is employed at high concentrations to obtain electrical contact for galvanic protection in environments between pH 5 and 10.

Table 7.1 shows the type and characteristics of these pigments.

In addition to lowering the cost, extenders also provide sag resistance to the liquid paint so that the edges remain covered. When the paint has dried they reduce the permeability to water and oxygen and provide reinforcing structure within the film. Talc and mica are used extensively as extenders. Mica is limited to approximately 10% of the total pigment. Both talc and mica, but particularly mica, reduce the permeability through the film as plate-like particles block permeation, forcing water and oxygen to seek a longer path through the binder around the particle.

K. Solvents

Organic solvents are usually required only to apply the coating and after application are designed to evaporate from the wet paint film. Water is considered either as a solvent or an emulsifier. The rate at which solvents evaporate influences the application characteristics of the coating. It is imperative that the solvent evaporate completely. If the solvents are partially retained and do not evaporate completely, premature failure by blistering or pinholing is likely to take place.

Most coatings are formulated to be applied at ambient conditions of approximately 75°F (24°C) and a 50% relative humidity. If ambient conditions are considerably higher or lower it is conceivable that solvent release could pose a problem.

L. Additives

Additives are used in coating formulations for specific purposes and in small, sometimes even trace amounts. Some typical additives and their purposes are as follows:

Phenylmercury, zinc, cuprous compounds: mildew inhibitors
Cobalt and manganese naphthanates: aid to surface and thorough drying
Zinc oxide: protection of the resin from heat and sun

Latex paints have a number of additives that act as stabilizers, coalescing aids, emulsion stabilizers, freeze–thaw stabilizers, and so on.

Table 7.1 Characteristics of Pigments for Metal Protective Paints

Pigment	Specific gravity	Color	Opacity	Special contribution to corrosion resistance
Active Pigments				
Red lead	8.8	Orange	Fair	Neutralizes film acids, insolubilizes sulfates and chlorides, renders water noncorrosive.
Basic silicon lead chromate	3.9	Orange	Poor	Neutralizes film acids, insolubilizes sulfates and chlorides, renders water noncorrosive.
Zinc yellow (chromate)	3.3	Yellow	Fair	Neutralizes film acids, anodic passivator, renders water noncorrosive
Zinc oxide (French process)	5.5	White		Neutralizes film acids, renders water noncorrosive
Zinc dust at low concentration in coatings for steel	7.1	Gray	Good	Neutralizes film acids
Galvanically Protective Pigments				
Zinc dust sacrificial at high concentration	7.1	Gray	Good	Makes electrical contact, galvanically sacrificial
Barrier Pigments				
Quartz	2.6	Nil	Translucent	Inert, compatible with vinyl ester additives
Extenders				
Mica	2.8	Nil	Translucent	Impermeability and inertness
Talc	2.8	Nil	Translucent	Impermeability and inertness
Asbestine	2.8	Nil	Translucent	Impermeability and inertness
Barytes	4.1	Nil	Translucent	Impermeability and inertness
Colorants				
Iron oxide	4.1	Red		Impermeability and inertness
Iron oxide	4.1	Ochre		Impermeability and inertness
Iron oxide	4.1	Black		Impermeability and inertness
Titanium dioxide	4.1	White	Excellent	Impermeability and inertness
Carbon black	1.8	Black	Good	Impermeability and inertness

Note: Titanium dioxide has better ''hiding'' than any other pigment.
Source: Ref. 1.

II. ZINC-RICH COATINGS

In all of the above-mentioned coatings the final film properties of corrosion protection and environmental resistance have been attributed to organic composition of their constituent resin or binder. The pigment, although playing an important role in the final film properties, is secondary to the resistance of the organic binders.

Conversely, in a zinc-rich coating, the role of the zinc pigment predominates. The high amount of zinc dust metal in the dried film determines the coating's basic property, i.e., galvanic protection. For a coating to be considered zinc-rich, the common rule of thumb is that there must be at least 75% by weight of zinc dust in the dry film. This may change since conductive extenders (notably di-iron phosphide) have been added to improve weldability and burn-through, with supposedly equivalent protection at lower zinc loadings.

The primary advantage of zinc-rich coatings is their ability to protect galvanically. The zinc pigment in the coating preferentially sacrifices itself in the electrochemical corrosion reaction to protect the underlying steel. This galvanic reaction, together with the filling and sealing effect of zinc reaction products (primarily zinc carbonate, zinc hydroxide, and complex zinc salts), provides more effective corrosion protection to steel substrates than does any other type of coating. Zinc-rich coatings cannot be used outside the pH range of 6–10.5.

III. FACTORS AFFECTING LIFE OF A PAINT COATING

For a paint film to perform effectively it must have mechanical resistance to external forces and resistance against environmental factors. As discussed, a paint film is a composite layer that is composed of a vehicle (resin) pigment and additives, and tends to contain defects in its structure. The common defects existing in a resultant paint film are:

1. Nonuniformity of vehicle molecules
2. Solvent residue
3. Residual stress
4. Differences in expansion coefficients in multilayer coating systems
5. Poor adhesion of polymer to pigment

A. Strength of Paint Film

Paint films require hardness, flexibility, brittleness resistance, abrasion resistance, mar resistance, and sag resistance. Paint coatings are formulated providing a balance of these mechanical properties. The mechanical strength of a paint film is described by the words "hardness and plasticity," which correspond to the modulus of elasticity and to the elongation at break obtained from the stress–strain curve of a paint film. Typical paint films have tensile properties as follows:

Paints	Tensile strength (g/mm)	Elongation at break (%)
Linseed oil	14–492	2–40
Alkyd resin varnish (16% PA)	141–1266	30–50
Amino-alkyd resin varnish (A/U = 7/3)	2180–2602	—
NC lacquer	844–2672	2–28
Methyl-*n*-butyl-*meta*-acrylic resin	1758–2532	19–49

Mechanical properties of paint coatings vary depending on the type of pigment, baking temperatures, and aging times. As baking temperatures rise, the curing of paint films is promoted and elongation reduced. Tensile strength is improved by curing and the elongation at breaks is reduced with increased drying time.

Structural defects in a paint film cause failures that are determined by environmental conditions such as thermal reaction, oxidation, photo-oxidation, and photothermal reaction. An important factor is controlling the physical properties of a paint film is the glass transition temperature (T_g). In the temperature range higher than T_g the motion of the resin molecules become more active, so that hardness, plasticity, and the permeabilities of water and oxygen vary greatly above the glass transition temperature. Table 7.2 lists the glass transition temperatures of organic films.

Deterioration of paint films is promoted by photolysis, photo-oxidation, or photothermal reaction as a result of exposure to natural light. Specifically, UV light (wavelength 40–400 nm) decomposes some polymer structures. Polymer films, such as vinyl chloride resins, are gradually decomposed by absorbing the energy of UV light.

The T_g of a polymer is of critical importance in the photolysis process. Radicals formed by photolysis are trapped in the matrix, but they diffuse and react at higher than T_g. The principal chains of polymers with keton groups form radicals:

Table 7.2 Glass Transition Temperatures of Organic Films

Organic film	Glass transition temp. (T_g, °C)
Phthalic acid resin	50
Acrylic lacquer	80–90
Chlorinated rubber	50
Bake-type melamine resin	90–100
Anionic resin	80
Cationic resin	120
Epoxy resin	80
Tar-epoxy resin	70
Polyurethane resin	40–60
Unsaturated polyester	80–90
Acrylic powder paint	100

$$RCOR' \rightarrow R + COR' \rightarrow CO^{\uparrow} + R'$$

$$ROCOR' \rightarrow R + OCOR' \rightarrow CO_2^{\uparrow} + R'$$

The resultant radicals accelerate the degradation of the polymer and in some cases HCl (polyvinyl chloride) or CH_4 is produced.

B. Adhesion of Paint Film

Paint-coated materials are exposed primarily to the natural atmosphere. Consequently, paint adhesion is influenced by atmospheric factors, particularly moisture and water. All organic polymers and organic coatings are permeable to water. They differ only in the degree of permeability. Table 7.3 provides typical values for the permeability and the diffusion coefficient for water in a number of different polymers.

When moisture permeates into the paint substrate interface, blisters form. There are four mechanisms whereby blisters can form:

1. Volume expansion due to swelling
2. Gas inclusion or gas formation
3. Osmotic processes due to soluble impurities at the film–substrate interface
4. Electroosmotic blistering

Of the four mechanisms, osmotic blister formation is the most important. Past experience has indicated that 70% of all paint coating failures are the

Table 7.3 Diffusion Data for Water Through Organic Films

Polymer	Temp. (°C)	p (× 10^9) [cm^3 (STP) cm] [cm^2/s cm Hg]	D(× 10^9) (cm^2/s)
Epoxy	25	10–44	2–8
	40	—	5
Phenolic	25	166	0.2–10
Polyethylene (low-density)	25	9	230
Polymethyl methacrylate	50	250	130
Polyisobutylene	30	7–22	—
Polystyrene	25	97	—
Polyvinyl acetate	40	600	150
Polyvinyl chloride	30	15	16
Vinylidene chloride/ acrylonitrile copolymer	25	1.7	0.32

Source: Ref. 4.

result of poor or inadequate surface preparation prior to application of the coating. Types of osmotically active surfaces at film–substrate interfaces are impurities such as sand particles from incomplete washing after wet sanding, water-soluble salt residues from phosphating, or sulfate nests in a rust layer. It is also possible for entrapped hydrophilic solvents to create blisters.

Regardless of the specific driving force, the blister formation process is shown in Figure 7.5 using an example of an intact coating on steel. When the coating is exposed to an aqueous solution, water vapor molecules, and, to a lesser extent, oxygen, they diffuse into the film toward the coating–substrate interface. Eventually, water may accumulate at the interface forming a thin water film of at least several monolayers. The accumulation takes place at sites of poor adhesion or at sites where wet adhesion problems arise. The degree to which permeated water may change the adhesion properties of a coated system is referred to as wet adhesion. No initial poorly adhered spots need be present for chemical disbondment to take place. A corrosion reaction can be initiated by the presence of an aqueous electrolyte with an electrochemical double layer, a cathodic species (oxygen) and an anodic species (metal). When the corrosion reaction has started at the interface the corrosion products can be responsible for the buildup of osmotic pressure, resulting in the formation of macroscopic blisters (see Figure 7.1B). Depending on the specific materials and circumstances the blisters may grow out because of the hydrodynamic pressure in combination with one of the

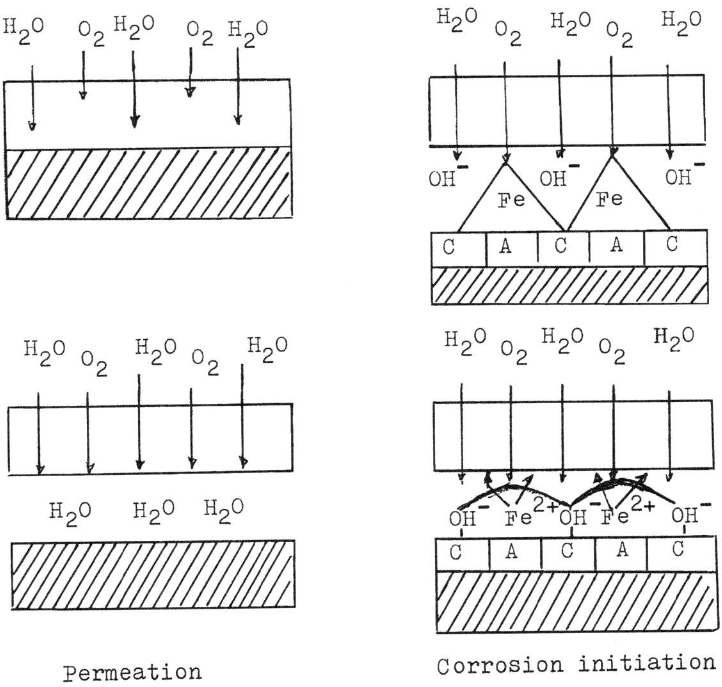

Figure 7.5 Initiation of a blister under an intact coating.

chemical propagation mechanisms such as cathodic delamination and anodic undercoating.

Oxygen migration through a paint film coating is generally much lower than that of water. In some instances the flow of oxygen increases with the water content of the film, probably because the water acts to swell the polymer. Table 7.4 gives the data for oxygen flow through free films of paint.

IV. SURFACE PREPARATION

Proper surface preparation is essential if the paint to be applied is to attain its maximum corrosion protection benefits. The specific preparation system to be specified will be dependent on the coating system to be applied and should usually be a "standard" surface preparation method. The most common methods are those defined by the Steel Structures Painting Council, which are shown in Table 7.5.

Table 7.4 Flux of Oxygen Through Representative Free Films of Paint, 100 μm Thick

Paint	J (mg/cm^2 day)
Alkyd (15% PVC Fe_2O_3)	0.0069
Alkyd (35% PVC Fe_2O_3)	0.0081
Alkyd-melamine	0.001
Chlorinated rubber (35% PVC Fe_2O_3)	0.017
Cellulose acetate	0.026 (RH = 95%)
Cellulose nitrate	0.115 (RH = 95%)
Epoxy melamine	0.008
Epoxy coal tar	0.0041
Epoxy-polyamide (35% PVC Fe_2O_3)	0.0064
Vinyl chloride/vinyl acetate copolymer	0.004 (RH = 95%)

Source: Ref. 4.

As important as the surface preparation is proper application of the paint. It is better to have a less effective paint applied properly than to have the best paint applied poorly.

V. ATMOSPHERIC EFFECTS AND COATING FAILURES

As mentioned previously, approximately 70% of all coating failures are the result of poor or inadequate surface preparation and/or application of the paint to the substrate. Typical coating failures include:

1. *Cracking, checking, and alligatoring*. These types of failures develop with the aging of the paint film. Shrinkage within the film during aging causes cracking and checking. Alligatoring is a film rupture, usually caused by application of a hard brittle film over a more flexible film.
2. *Peeling, flaking, and delamination*. These failures are caused by poor adhesion. When peeling or flaking occurs between coats it is called delamination.
3. *Rusting*. Failure of a coated surface may appear as (1) spot rusting at minute areas, (2) pinhole rusting at minute areas, (3) rust nodules breaking through the coating, and (4) underfilm rusting, which eventually causes peeling and flaking of the coating.
4. *Lifting and wrinkling*. When the solvent of a succeeding coat of paint too rapidly softens the previous coat, lifting results. Rapid

Table 7.5 Summary of Surface Preparation Specifications

SSPC specification	Description
SP 1: Solvent cleaning	Removal of oil, grease, dirt, soil, salts, and contaminants by cleaning with solvent, vapor, alkali, emulsion, or steam
SP 2: Hand-tool cleaning	Removal of loose rust, mill scale, and paint, to degree specified, by hand chipping and wire brushing
SP 3: Power-tool cleaning	Removal of loose rust, mill scale, and paint, to degree specified, by power tool, chipping, descaling, sanding, wire brushing, and grinding
SP 5: White-metal blast cleaning	Removal of all visible rust, mill scale, paint, and foreign matter by blast cleaning, by wheel or nozzle, dry or wet, using sand, grit, or shot for very corrosive atmospheres where high cost of cleaning is warranted
SP 6: Commercial blast cleaning	Blast cleaning until at least two-thirds of the surface area is free of all visible residues (for severe exposure)
SP 7: Brush-off blast cleaning	Blast cleaning of all except tightly adhering residues of mill scale, rust, and coatings, exposing numerous evenly distributed flecks of underlying metal
SP 8: Pickling	Complete removal of rust and mill scale by acid pickling, duplex pickling, or electrolytic pickling
SP 10: Near-white blast cleaning	Blast cleaning to near–white metal cleanliness until at least 95% of the surface area is free of all visible residue (for high-humidity, chemical atmosphere, marine, and other corrosive environments)

surface drying of a coating without uniform drying throughout the rest of the film results in a phenomenon known as wrinkling.

5. *Failures around weld areas*. Coating adhesion can be hampered by weld flux, which may also accelerate corrosion under the film. Relatively large projections of weld spatter cause possible gaps and cavities that may not be coated sufficiently to provide protection.
6. *Edge failures*. Edge failures usually take the form of rusting through the film at the edge where the coating is normally the thinnest. This is normally followed by eventual rust creepage under the film.

Factors in the atmosphere that cause corrosion or degradation of the coating are moisture, water, UV light, temperature, oxygen, ozone, pollutants, and wind. Types of failures caused by these factors are as follows:

1. *Chalking*. UV light, moisture, oxygen, and chemicals degrade the coating resulting in chalk.
2. *Color fading or color change*. Color fading or color change may be caused by chalk on the surface or by breakdown of the colored pigments. Pigments can be decomposed or degraded by UV light or reaction with chemicals.
3. *Blistering*. Blistering may be caused by:
   a. Moisture vapor that passes through the film and condenses at a point of low adhesion
   b. Inadequate release of solvent during both application and drying of the coating system
   c. Poor surface preparation
   d. Poor adhesion of the coating to the substrate or poor intercoat adhesion
   e. A coat within the paint system that is not resistant to the environment
   f. Application of a relatively fast-drying coating over a relatively porous surface
   g. Either direct or stray cathodic protection currents
4. Failures due to chemical or solvent attack. When a coating is not resistant to its chemical or solvent environment there is apparent disintegration of the film.

Any atmospheric environment is subject to dry and wet cycles. Since water and moisture have a decided effect on the degradation of a coating, the time of wetness of a coating is important. Moisture and water that attack organic films are from the rain, fog, dew, snow, and water vapors in the

Table 7.6 Relationship of Bond Strength to Exposure Time in 100% Relative Humidity

	Bond strength (psi)		
Exposure time (hr)	Epoxy ester	Polyurethane	Thermosetting acrylic
Initial	4790	3410	5700
24	1640	1500	3650
48	1500	1430	3420
120	—	1390	2400
195	1400	1130	1850
500	1390	670	480

atmosphere. Relative humidity is a particularly important factor. As exposure time is increased in 100% relative humidity, the bond strength of a paint coating is reduced. This is shown in Table 7.6.

Temperature fluctuations and longer time of wetness tends to produce clustered water, which increases the acceleration of the degradation of the organic film, particularly in a marine atmosphere. The most severe atmosphere for a paint film is that of a seashore environment. In addition to time of wetness, sunlight—particularly the UV rays of short wavelengths—promotes the degradation of paint coatings. Chalking and breaking down of colored pigment are accelerated, causing color change or fading.

VI. APPLICATIONS

Long service life is required of paint coating systems for steel structures and other items. In general, paint coating systems for steel structures are classified as general and heavy coatings.

General paint coating systems are a combination of oil-type rust preventative primer and long oil alkyd resin paint or chlorinated rubber paint (intermediate and overcoats) with a total thickness of less than 150 μm. These are applied in relatively mild atmospheres such as rural, mountain, and city.

Heavy paint coating systems are used in severe atmospheres such as marine and industrial, and consist of a spray coating of zinc or aluminum, or a zinc-rich paint primer, followed by several layers of synthetic resin paints such as epoxy, tar epoxy, or polyurethane resins. The total film thickness is more than 250 μm.

Types of paint coatings for use in various environments are as follows:

1. *Rural and mountain atmospheres.* These atmospheres are usually mild. An oil-type rust preventative primer and phthalic acid resin paint overcoat is used. Chlorinated rubber and epoxy paints are used in areas having high humidity and heavy snowfalls.
2. *Urban atmospheres.* Chlorinated rubber and epoxy paints are suitable for these atmospheres, which are usually polluted by automobile exhaust and waste gases from factories.
3. *Industrial atmospheres.* Epoxy and vinyl resin paints are recommended for these atmospheres, which contain acid and alkaline gases and chemicals.
4. *Marine atmospheres.* Salt particles are found in these atmospheres. Because of their salt spray resistance, chlorinated rubber, epoxy, and vinyl resin paint coatings are used.

5. *Acid atmospheres*. Some industrial or marine atmospheres may also have an abnormally high acidity. Under these conditions phenolics, epoxy ester, chlorinated rubber, chlorinated vinyl, epoxy polyurethane, and tar-epoxy resin paints are recommended.
6. *Alkaline atmospheres*. Some industrial atmospheres may have an abnormally high alkalinity. Under these conditions epoxy ester, chlorinated rubber, chlorinated vinyl, epoxy, polyurethane, and tar-epoxy paints are recommended.
7. *Solvent atmospheres*. Epoxy, polyurethane, and tar-epoxy paint coatings are used in areas having a high solvent concentration in the atmosphere.

VII. SUMMARY

The effectiveness of a paint coating to provide protection to the substrate is dependent on three factors: coating selection, surface preparation, and application (including proper curing). First, it is essential that the proper coating system be selected to combat the corrosiveness of the atmosphere. Second, the surface of the substrate must be properly prepared to receive the coating. Third, the coating must be properly applied including proper coating thickness, and sufficient time allowed for curing. If these steps are followed, a durable coating will be achieved.

REFERENCES

1. KB Tator. Coatings. In: PA Schweitzer, ed. Corrosion and Corrosion Protection Handbook, 2nd ed. New York: Marcel Dekker, pp. 453–489.
2. I Suzuki. Corrosion Resistant Technologies. New York: Marcel Dekker, 1989.
3. WM McMahon. Painting for Protection. In: PA Schweitzer, ed. Corrosion Engineering Handbook. New York: Marcel Dekker, 1996, pp. 563–580.
4. H Leidheiser Jr. Coatings. In: F Mansfield, ed. Corrosion Mechanisms. New York: Marcel Dekker, 1987, pp 165–209.
5. JHW deWit. Inorganic and organic coatings. In: P Marcus, J Oudar, eds. Corrosion Mechanisms in Theory and Practice. New York: Marcel Dekker, 1995, pp. 581–628.

8
Atmospheric Corrosion of Plastics

Polymers are the building blocks of plastics. The term is derived from the Greek meaning "many parts." They are large molecules composed of many repeat units that have been chemically bonded into long chains. Wool, silk, and cotton are examples of natural polymers.

The monomer building blocks are chemically bonded by a process known as polymerization, which can take place in one of several methods. In condensation polymerization the reaction between monomer units, or chain-end groups, releases a small molecule, often water. This is an equilibrium reaction that will halt unless the byproduct is removed. Plastics produced by this process will degrade when exposed to water and high temperatures. In addition, polymerization, a chain reaction, apends new monomer units to the growing polymer molecule one at a time. Each new unit creates an active site for the next attachment.

Over the years the chemical industry has developed a wide range of synthetic polymers for a diversity of products. Celluloid, developed by Hyatt in 1868, was the forerunner of plastics. Bakelite, a phenol-formaldehyde resin, followed in 1907, nylon in 1938, the polyesters and polyethylene in 1942, the epoxies and acrylonitrile-butadiene-styrene (ABS) in 1947. Large-scale usage of these materials developed in the 1950s.

Plastics are composed primarily of carbon, hydrogen, chlorine, fluorine, oxygen, silicon, and nitrogen in various combinations. The elements are held together with single, double, and triple bonds in both aliphatic and aromatic structures. Because of the various combinations and arrangements a wide variation of properties exist between the different plastics. These properties can be modified by the addition of fibers, fillers, ultraviolet (UV) light stabilizers, colored pigments, and flame retardants. Specific properties may also be achieved by the blending of different polymers.

Plastics that are filled and reinforced are referred to as composites or laminates. Laminates are formed by using multiple layers of fiber-reinforced resin or different layers of different polymers.

There are two distinct groups of plastics—the thermoplasts and the thermosets. The thermoplasts can be melted, cooled, and remelted without destroying the physical or mechanical properties of the polymer. The thermosets begin as powder or liquid that is reacted with a second material, or which through catalyzed polymerization results in a new product, which has properties differing from those of either starting material.

I. ATMOSPHERIC CORROSION OF PLASTICS

Atmospheric corrosion has been defined as an electrochemical reaction requiring the presence of an electrolyte. This is true of a metallic material. However, polymers do not corrode by electrochemical mechanisms as do metals and alloys. The degradation of the plastic or its properties is related to the structured similarity between the material and the environment and/or the permeability of the material. Permeability is controlled by the solubility of the material in the environment and by the rate of diffusion of the environment into the plastic. When the diffused environment is chemically incompatible with the plastic matrix, all of the mechanical properties of the plastic are reduced.

Plastics are attacked by chemical reaction or solvation. Solvation is the penetration of the polymer by a corrodent, which can cause softening, swelling, and ultimate failure. Corrosion of plastics can be classified in the following ways as to attack mechanism.

1. Disintegration or degradation of a physical nature due to absorption, permeation, solvent action, or other factors.
2. Oxidation, where chemical bonds are attacked.
3. Hydrolysis, where ester linkages are attacked.
4. Radiation.
5. Thermal degradation involving depolymerization and possibly repolymerization.
6. Dehydration (somewhat uncommon).
7. Any combination of the above.

Results of such attacks will appear in the form of softening, charring, crazing, delamination, embrittlement, discoloration, dissolution, or swelling.

The corrosion resistance of polymer matrix composites is also affected by two other factors: the nature of the laminate and, in the case of thermoset resins, the cure. Improper or insufficient cure time will adversely affect the corrosion resistance.

Polymeric materials in outdoor applications are exposed to weather extremes that can be extremely deleterious to the material. The most harmful weather component, exposure to UV radiation can cause embrittlement, fading, surface cracking, and chalking. Most plastics, after being exposed in direct sunlight for a period of years, exhibit reduced impact resistance, lower overall mechanical performance, and a change in appearance.

The electromagnetic energy from sunlight is normally divided into UV light, visible light, and infrared energy. Infrared energy consists of wavelengths longer than the visible red wavelengths and starts above 760 nm. Visible light is defined as radiation between 400 and 760 nm. Ultraviolet light consists of radiation below 400 nm. The UV portion of the spectrum is further subdivided into UV-A, UV-B, and UV-C. The effects of the various wavelength regions are shown in Table 8.1.

Since UV is easily filtered by air masses, cloud cover, pollution, and other factors, the amount and spectrum of natural UV exposure is extremely variable. Because the sun is lower in the sky during the winter months it is filtered through a greater air mass. This creates two important differences between summer and winter sunlight: changes in the intensity of the light and in the spectrum. During the winter months much of the damaging short-wavelength UV light is filtered out. For example, the intensity of UV at 320 nm changes about 8 to 1 from summer to winter. In addition, the short-wavelength solar cut-off shifts from about 295 nm in summer to about 310 nm in winter. As a result, materials sensitive to UV below 320 nm would degrade only slightly, if at all, during the winter months.

Table 8.1 Wavelength Regions of the UV

Region	Wavelength (nm)	Characteristics
UV-A	400–315	Causes polymer damage.
UV-B	315–280	Includes the shortest wavelengths at the earth's surface. Causes severe polymer damage. Absorbed by window glass.
UV-C	280–100	Filtered by earth's atmosphere, found only in outer space.

Photochemical degradation is caused by photons of light breaking chemical bonds. For each type of chemical bond, there is a critical threshold wavelength of light with enough energy to cause a reaction. Light of any wavelength shorter than the threshold can break the bond, but longer wavelengths of light cannot break it, regardless of their intensity (brightness). Therefore, the short-wavelength cutoff of a light source is of critical importance. If a particular polymer is sensitive only to UV light below 295 nm (the solar cutoff point) it will never experience photochemical deterioration outdoors.

The ability to withstand weathering varies with the polymer type and within grades of a particular resin. Many resin grades are available with UV-absorbing additives to improve weatherability. However, the higher molecular weight grades of a resin generally exhibit better weatherability than lower molecular weight grades with comparable additives. In addition, some colors tend to weather better than others.

II. THERMOPLASTS

The periodic table provides a means whereby the general difference in the corrosion resistance of the thermoplasts can be determined. In the periodic table the basic elements of nature are placed into classes with similar properties, i.e., elements and compounds that exhibit similar behavior. These classes are alkali metals, alkaline earth metals, transition metals, rare earth series, other metals, nonmetals, and noble (inert) gases.

Of particular importance and interest in the case of thermoplasts is the category known as halogens. The elements included in this category are fluorine, chlorine, bromine, and iodine. Since these are the most electronegative elements in the periodic table they are the most likely to attract an electron from another element and become a stable structure. Of all of the halogens, fluorine is the most electronegative, permitting it to bond strongly with carbon and hydrogen atoms, but not well with itself. The carbon–fluorine bond is predominant in polyvinylidine fluoride (PVDF) and is responsible for the important properties of these materials. These are among the strongest known organic compounds. The fluorine acts as a protective shield for other bonds of lesser strength within the main chain of the polymer. The carbon–hydrogen bond of which such plastics as polyethylene (PE) and polypropylene (PP) are composed is considerably weaker. The carbon–chlorine bond, a key bond in polyvinyl chloride (PVC), is still weaker.

The arrangement of the elements in the molecule, the symmetry of the structure, and the degree of branching of the polymer chains are as important as the specific elements contained in the molecule. Plastics containing the carbon–hydrogen bonds, such as PP and PE, and the carbon–chlorine bonds, such as PVC, ethylene trifluorethylene (ECTFE), and chlorotrifluorethylene (CTFE) are different in the important property of chemical resistance from a fully fluorinated plastic such as polytetrafluorethylene (PTFE). Table 8.2 lists the abbreviations used for the thermoplasts. In considering these polymers for outdoor use, the surrounding atmospheric temperatures must be considered, as well as the operating temperatures. Certain geographic areas may have ambient temperatures at times that will be outside the allowable ranges of some polymers.

A relative new series of polymers have been developed known as "engineering plastics," "engineering polymers," or "high-performance plastics." Regardless of which term is used they all refer to those higher strength

Table 8.2 Abbreviations Used for Thermoplasts

ABS	acrylonitrile-butadiene-styrene
CPVC	chlorinated polyvinyl chloride
CTFE	chlorotrifluorethylene
ECTFE	ethylene chlorotrifluorethylene
ETFE	ethylene tetrafluorethylene
FEP	fluorinated ethylene propylene
HDPE	high-density polyethylene
PA	polyamide
PAI	polyamideimide
PB	polybutylene
PC	polycarbonate
PCTFE	polychlorotrifluorethylene
PE	polyethylene
PEEK	polyetheretherketone
PEI	polyetherimide
PEK	polyetherketone
PES	polyethersulfone
PF	phenol formaldehyde
PFA	polyfluoralkoxy
PI	polyimide
PP	polypropylene
PPS	polyphenylene sulfide
PSF	polysulfone
PTFE	polytetrafluorethylene
PVC	polyvinyl chloride
PVDC	polyvinylidene chloride
PVDF	polyvinylidene fluoride
UHMWPE	ultrahigh molecular weight polyethylene

and better high-temperature resistance polymers to distinguish them from the general-purpose polymers. These engineering thermoplastics lend themselves for use in engineering design, competing in terms of load bearing characteristics with metals and ceramics. Applications include gears and structural members.

The general-purpose thermoplastic polymers consist of such materials as polyethylene, polypropylene, polyurethanes, and polyvinyl chlorides. The engineering thermoplastic polymers include the polyimides, polyamideimide, polysulfones, polyetheretherketones, and polyphenylene sulfides.

A. Acrylonitrile-Butadiene-Styrene

ABS plastics are derived from acrylonitrile, butadiene, and styrene with the following general chemical structure:

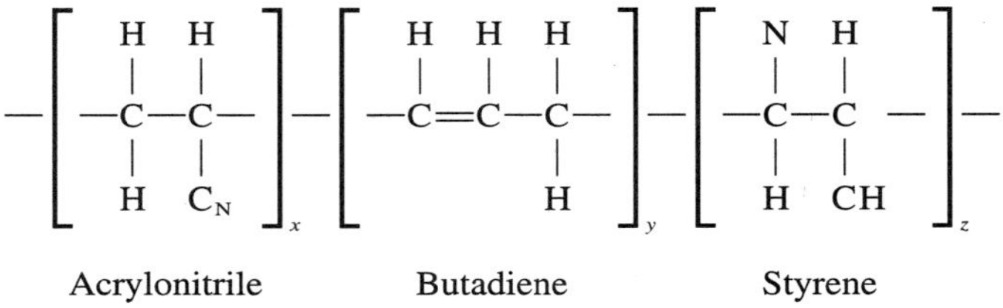

In general, this class of polymers possesses hardness and rigidity without brittleness, and has a good balance of tensile strength, impact resistance, and low-temperature properties. ABS will be degraded by UV light unless protective additives are incorporated into the formulation. It has an operating temperature range of −40 to 140°F (−40 to 60°C).

B. Chlorinated Polyvinyl Chloride

Acetylene and hydrochloric acid are reacted together to produce PVC with a chlorination of approximately 56.5%. Further chlorination of the PVC to approximately 67% produces chlorinated PVC (CPVC) with the following chemical structure:

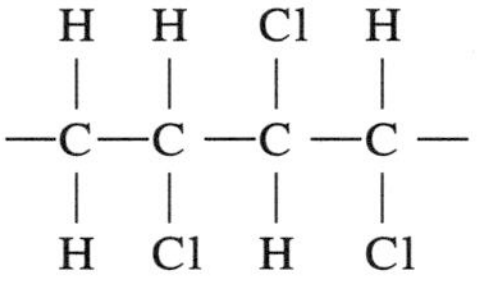

CPVC is resistant to most acids, alkalies, salts, halogens, sulfur dioxide, and nitrogen dioxide. It has an operating temperature range of 0–180°F (−18 to 82°C).

C. Ethylene Chlorotrifluorethylene

Ethylene chlorotrifluorethylene (ECTFE) is a 1:1 alternating polymer of ethylene and chlorotrifluorethylene having the following chemical structure:

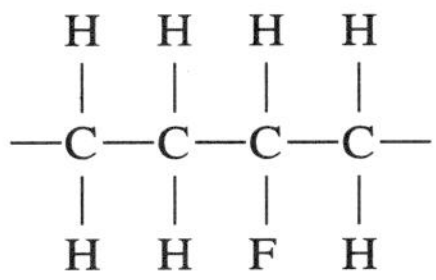

ECTFE has excellent chemical and abrasion resistance. It is unaffected by atmospheric pollutants and weathering, including UV radiation. ECTFE has an operating temperature range of −150 to 340°F (−76 to 171°C).

D. Ethylene Tetrafluorethylene

Ethylene tetrafluorethylene (ETFE) is a high-temperature fluoropolymer with the following structural formula:

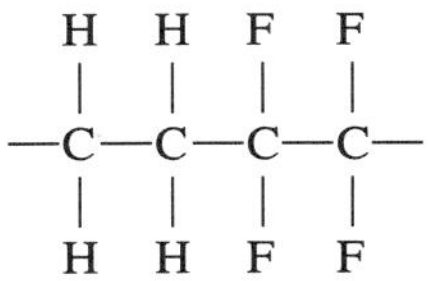

It is an alternating copolymer of ethylene and tetrafluorethylene sold by DuPont under the tradename Tefzel. ETFE is weather-resistant and UV-resistant with an operating temperature range of −370 to 300°F (−223 to 149°C). It is also inert to normal atmospheric pollutants.

E. Fluorinated Ethylene Propylene

Fluorinated ethylene propylene (FEP) is a fully fluorinated thermoplastic with some branching but consists mainly of linear chains having the following formula:

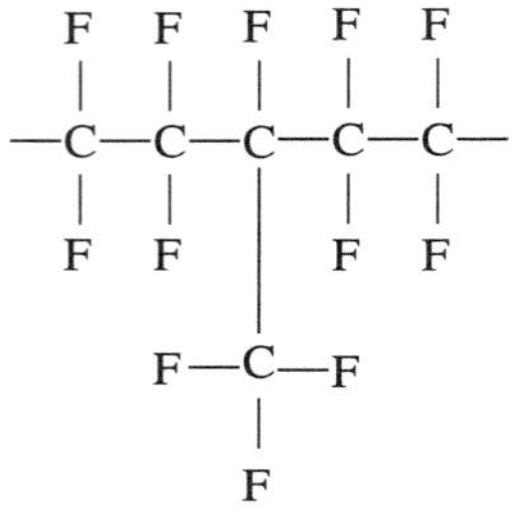

FEP has excellent chemical resistance, being resistant to all atmospheric pollutants. Except for some chlorinated hydrocarbons, it has excellent permeation resistance. It is not degraded by UV light and has otherwise excellent weathering resistance. FEP has an operating temperature range of −50 to 400°F (−45 to 205°C).

F. High-Density Polyethylene

High-density polyethylene (HDPE) is a member of the group of polymers known as polyolefins. It is produced by polymerizing ethylene gas obtained from petroleum hydrocarbons and is available in several types. The basic structure is:

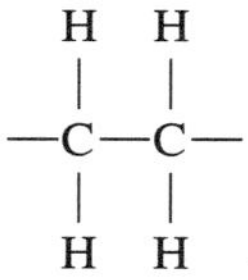

The various types of polyethylene differ in molecular structure, crystallinity, molecular weight, and molecular weight distribution.

The terms low, medium, and high density refer to the ASTM designation based on the unmodified PE. Low-density polyethylene has a specific gravity of 0.91–0.925, medium-density PE has a specific gravity of 0.926–0.940, and high-density PE has a specific gravity of 0.941–0.959. The densities, being related to the molecular structure, are indications of the properties of the final product.

There are two grades that are primarily used for corrosion-resistant applications: high-density polyethylene (HDPE) and the ultrahigh molecular weight polyethylene (UHMWPE). The HDPE has an average molecular weight of 200,000–500,000, whereas the UHMWPE has an average molecular weight of at least 3.1 million.

Polyethylene is resistant to a wide variety of materials including acids, bases, inorganic salts, and many fertilizer solutions. It is resistant to sulfur dioxide and nitrogen dioxide but does not possess good weathering properties. If exposed to UV radiation from the sun or another source, photo- or light oxidation will occur. To prevent this it is necessary to incorporate carbon black in the resin. Other stabilizers will not provide complete protection. PE will not support biological growth.

The primary difference between HDPE and UHMWPE is in the maximum allowable operating temperature, with the UHMWPE having an approximately 20°F higher allowable temperature. Operating temperature of HDPE is −60 to 180°F (−51 to 82°C), whereas the operating temperature for UHMWPE is −60 to 200°F (−51 to 93°C).

G. Polyamides

The polyamides (PAs) are commonly known as nylons. They are linear molecules with a high degree of crystallinity and have the following formulas, as shown in Figure 8.1. As can be seen from the chemical formulas, there are a variety of polyamides available. Shown are the varieties that are used for corrosion resistance. The nylons are resistant to UV degradation, weathering, and ozone. However, all nylons will absorb some moisture from environmental humidity. This can have an effect on the dimensional stability of a part.

Nylons 6, 11, and 6/6 are resistant to wet sulfur dioxide. Operating temperature for the polyamides is −60 to 300°F (−51 to 149°C).

H. Polyamideimide

Polyamideimides (PAIs) are heterocyclic polymers having a noncarbon atom of nitrogen in one of the rings of the molecular chain. A typical chemical structure is shown below:

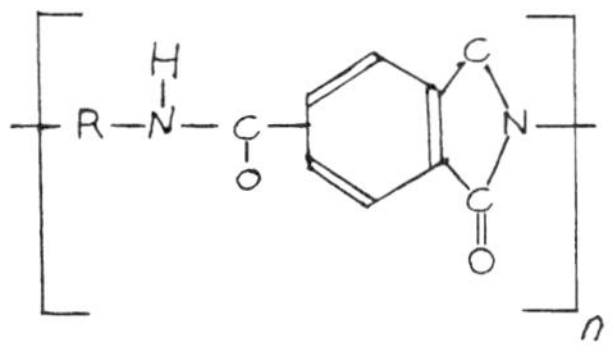

This series of thermoplasts are capable of use at high and low temperature and as such find application in the extreme environments of space. They also have resistance to UV degradation and very low outgassing in high

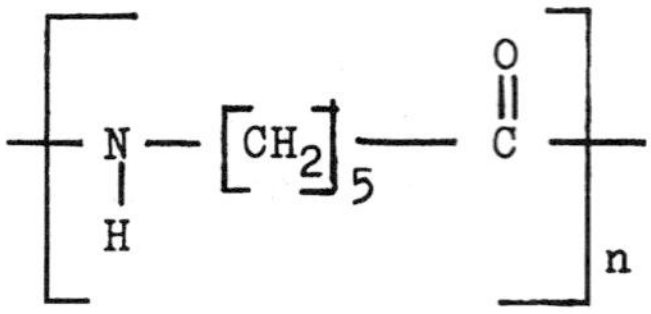

Nylon 6

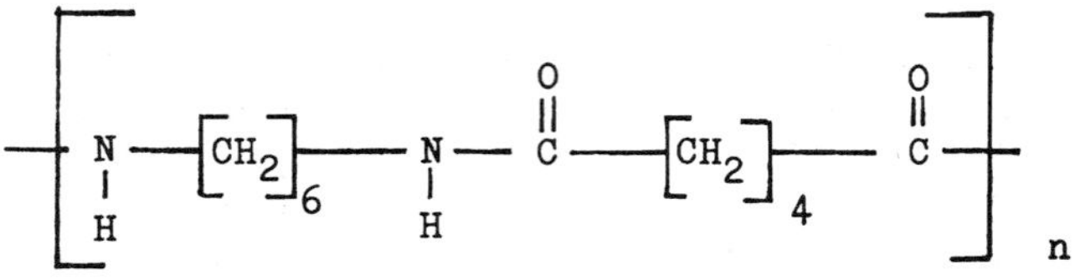

Nylon 6/6

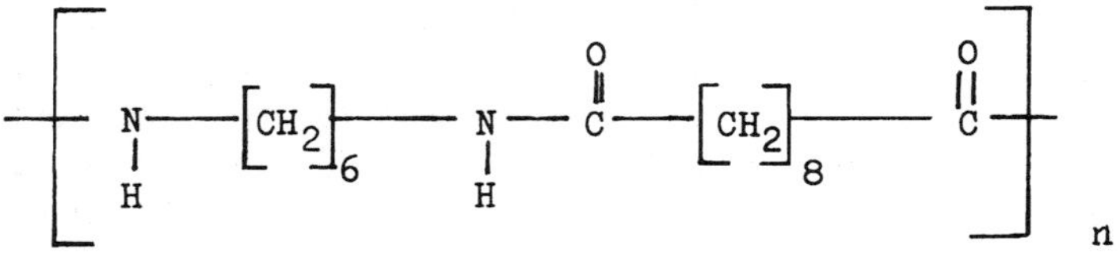

Nylon 6/10

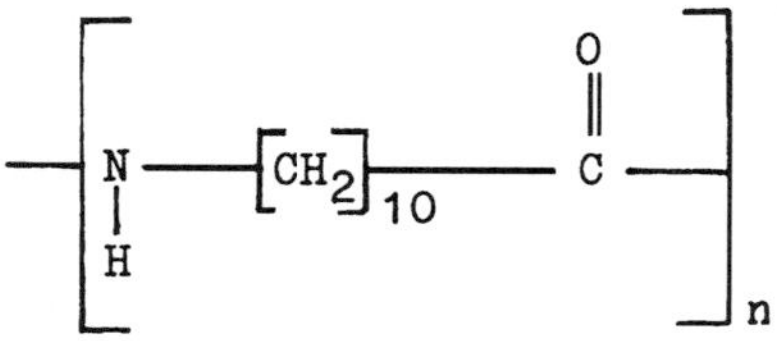

Nylon 11

Figure 8.1 Polyamide (nylon) formulas.

vacuum. The temperature range is from below −310°F to 500°F (below −190 to 250°C).

I. Polybutylene

Polybutylene (PB) is a member of the polyolefin family. It is resistant to acids, bases, soaps, and detergents but is not completely resistant to aliphatic solvents at room temperatures. PB is partially soluble in aromatic and chlorinated hydrocarbons above 140°F (60°C) and is subject to stress cracking with even low concentrations of chlorine. Operating temperature for PB is a maximum of 200°F (93°C).

J. Polycarbonate

Polycarbonate (PC) is one of the groups of plastics classified as an engineering thermoplastic because of its high performance in engineering designs. The generalized chemical structure is shown below:

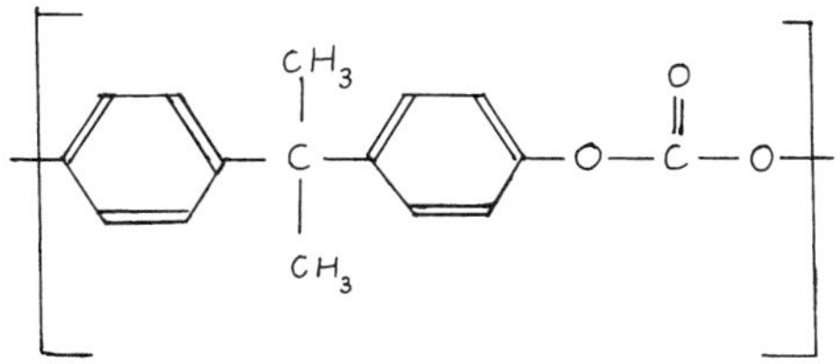

Polycarbonate is also produced under the tradename Lexan (General Electric Plastics). Because of its extremely high impact resistance and good clarity it is widely used for windows in chemical equipment and glazing in chemical plants. Because of its exceptional weatherability, corrosion resistance, and high-impact strength, it is highly useful in outdoor energy management devices, network interfaces, electrical wiring blocks, telephone equipment, and lighting diffusers, globes, and housings.

Immersion in water and exposure to high humidity at temperatures up to 212°F (100°C) have little effect on dimensions. Polycarbonates are among the most stable polymers in a wet environment. Operating temperature for polycarbonates is −200 to 250–275°F (−129 to 120–135°C).

K. Polychlorotrifluorethylene

Polychlorotrifluorethylene (PCTFE) is manufactured by the 3M Company and goes under the tradename of Kel-F. It has a structural formula as shown below:

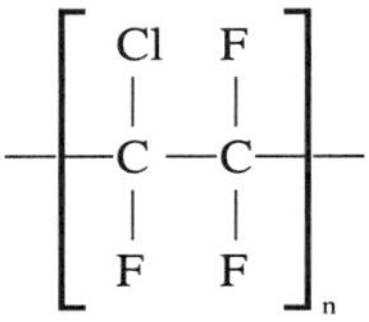

Within its service temperature range it exhibits greater tensile and compressive strength than PTFE, but has less chemical resistance than PTFE, FEP, or PFA. PCTFE has the lowest water transmission rate of any of the thermoplasts, and has good weathering properties. The operating temperatures range for PCTFE is −150 to 380°F (−76 to 190°C).

L. Polyetheretherketone

Polyetheretherketone (PEEK) is an engineering thermoplast having the following formula:

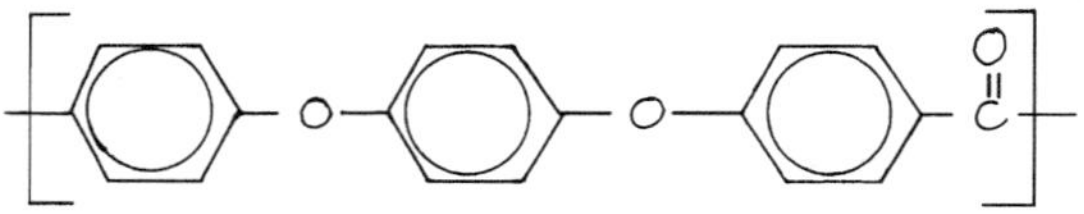

Like most polyaromatics, PEEK is subject to degradation by UV light during outdoor weathering. However, testing has shown that over a 12-month period for both natural and pigmented moldings the effect is minimal. In more extreme weathering conditions, painting or pigmenting of the polymer protects it from excessive property degradation.

PEEK is not chemically attacked by water. It has excellent long-term resistance to water at both ambient and elevated services. It also has excellent rain erosion resistance. This has led to a number of aircraft exterior applications such as radomes and fairings. PEEK has an operating temperature range of −85 to 480°F (−65 to 250°C).

M. Polyetherimide

Polyetherimide (PEI) is an engineering thermoplastic with the structural formula shown in Figure 8.2. The PEIs are alternating aromatic ether and imide

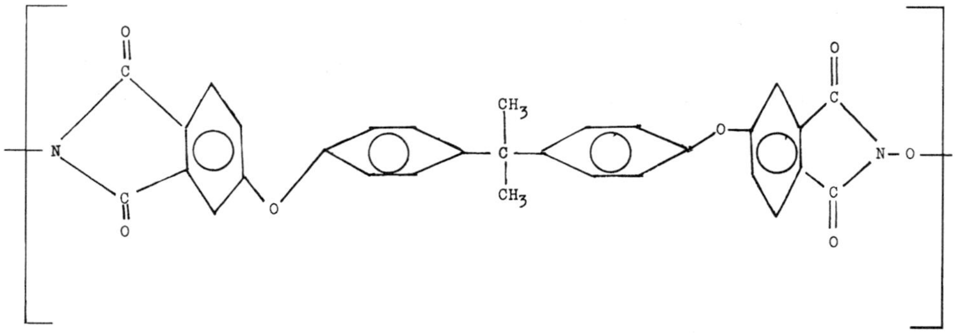

Figure 8.2 Polyetherimide structural formula.

units, making them one of the strongest thermoplastics. They have an operating temperature range of below −310 to 500°F (below −190 to 260°C). These polymers are used primarily for high-temperature applications.

N. Polyethersulfone

Polyethersulfone (PES) is an engineering thermoplastic with the following structural formula:

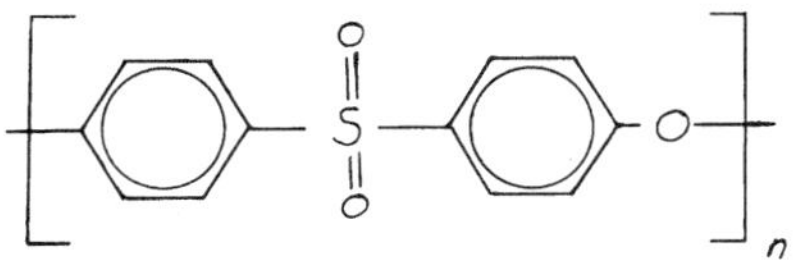

PES does not have good outdoor weathering properties. Since it is susceptible to degradation by UV light if used outdoors it must be stabilized by incorporating carbon black or by painting. It will also absorb small quantities of moisture from the atmosphere resulting in small dimensional changes. At equilibrium water content for 65% relative humidity the dimensional changes are of the order of magnitude of 0.15%. The maximum operating temperature is 300°F (170°C).

O. Perfluoralkoxy

Perfluoralkoxy (PFA) is a fully fluorinated polymer having the following formula:

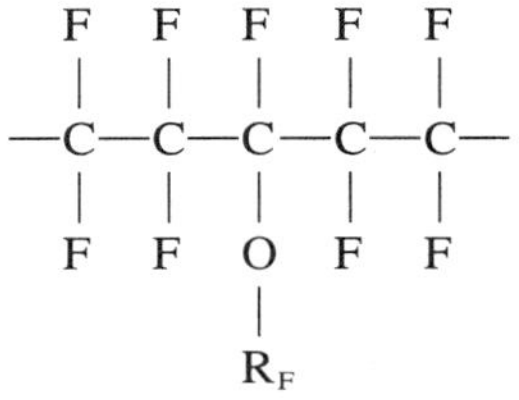

$$R_F = C_nF_{2n} + 1$$

PFA is inert to most chemicals, has excellent weatherability, and is not subject to UV degradation. It is subject to permeation by certain gases. The operating temperature range is from −310 to 500°F (−190 to 260°C).

P. Polyimides

The polyimides (PIs) are an engineering plastic with high heat resistance. They have the following structural formula:

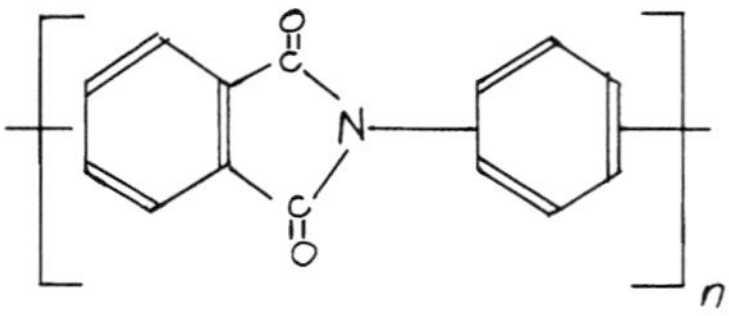

They have excellent chemical and radiation inertness and are not subject to UV degradation. High oxidation resistance is another important property. The operating temperature range is from below −310 to 500–600°F (below −190 to 260–315°C).

Q. Polypropylene

Polypropylene (PP) is a member of the olefin family. The polyolefins are composed only of carbon and hydrogen having the following structure:

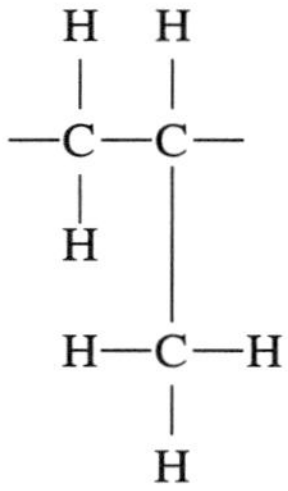

PP is available as a homopolymer or a copolymer with polyethylene. Since the homopolymer has superior chemical, thermal, and physical properties (except for impact resistance), it is preferred for difficult chemical, thermal, and physical conditions.

Because PP is subject to degradation by UV light a UV absorber or screening agent must be used to protect the material when exposed to sunlight. It has an operating temperature range of 32–215°F (0–102°C).

R. Polyphenylene Sulfide

Polyphenylene sulfide (PPS) has a rigid backbone chain consisting of recurring para-substituted benzene rings and sulfur atoms. The chemical structure is as follows:

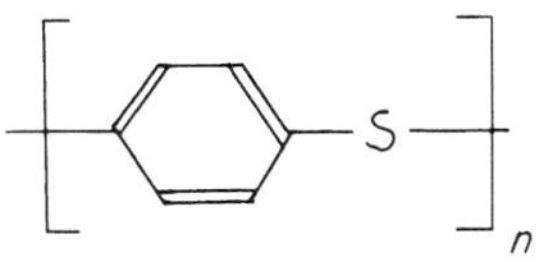

It is sold under the tradename of Ryton. PPS is resistant to aqueous inorganic salts and bases and can be used under highly oxidizing conditions. It has a maximum operating temperature of 450°F (230°C).

S. Polysulfone

Polysulfone (PSF) is an engineering plastic with the following chemical structure:

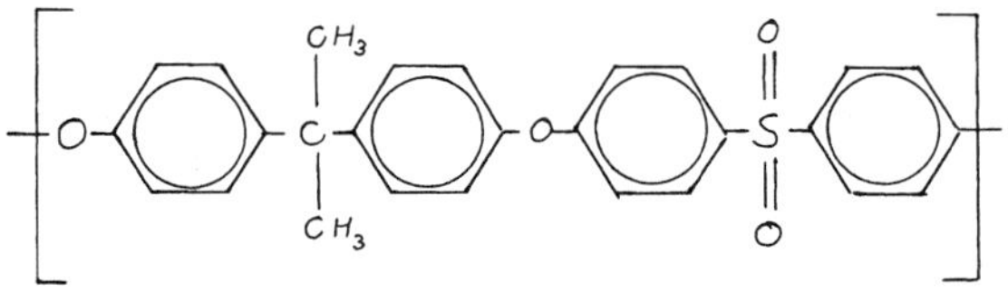

PSF resists hydrolysis and aqueous acid and alkaline environments. It also has good weatherability and an operating temperature range of −150 to 300°F (−101 to 149°C).

T. Polytetrafluorethylene

Polytetrafluorethylene (PTFE) is a fully fluorinated thermoplastic having the following formula:

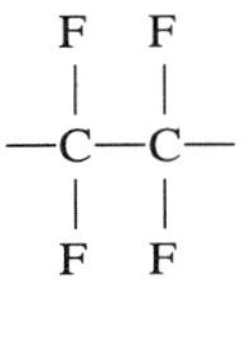

PTFE is resistant to all normal atmospheric pollutants, is not degraded by UV radiation, and has excellent weathering properties. It has an operating temperature range of −20 to 500°F (−29 to 260°C).

U. Polyvinyl Chloride

Polyvinyl chloride (PVC) is the most widely used of the thermoplasts. It is polymerized vinyl chloride, which is produced from acetylene and anhydrous hydrochloric acid. The structure is as follows:

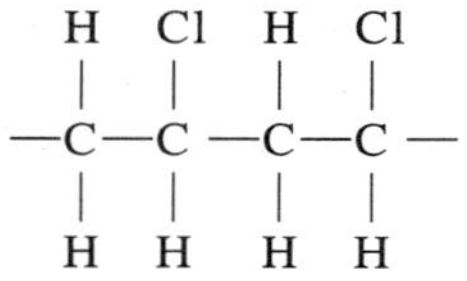

PVC is produced in types 1 and 2. Type 1 is an unplasticized form having good chemical resistance and normal impact. Type 2 has been modified by the addition of styrene-butadiene rubber to improve notch toughness and impact strength. By so doing the chemical resistance has been reduced.

Type 1 PVC resists attack by most acids and strong alkalies, gasoline, kerosene, aliphatic alcohols, and hydrocarbons. The chemical resistance of type 2 PVC to oxidizing and highly alkaline materials is reduced.

PVC is resistant to all of the normal atmospheric corrodents including weather and UV degradation. The primary applications are for piping and electrical conduit and wire insulation. It may be colored.

PVC has an operating temperature range of 0 to 140°F (−18 to 60°C).

V. Polyvinylidene Chloride

Polyvinylidene chloride (PVDC) is also known under the tradename of Saran. It is a variant of PVC having both chlorine atoms at the same end of the monomer instead of at the opposite ends as shown below:

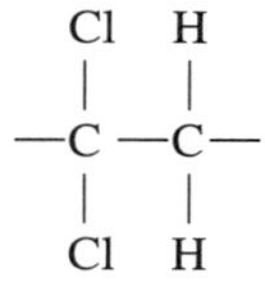

PVDC has improved strength, hardness, and chemical resistance, particularly to oxidants, mineral acids, and solvents, over that of PVC. It is resistant to all of the normal atmospheric pollutants including weather and UV degradation. Applications include piping and equipment components (pumps, valves, etc.). The operating temperature range is 0 to 175°F (−18 to 80°C).

W. Polyvinylidene Fluoride

Polyvinylidene fluoride (PVDF) has a chemical structure similar to that of PTFE except that it is not fully fluorinated. The chemical structure is shown below:

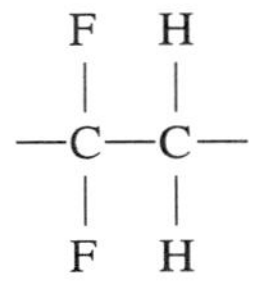

PVDF is chemically resistant to most acids, bases, and organic solvents. It is also resistant to all of the normal atmospheric pollutants and has excellent resistance to UV degradation. Applications are found in piping and electrical insulation. PVDF has an operating temperature range of −50 to 275°F (−45 to 135°C).

X. Summary

It must be remembered that all thermoplastic polymers are compounded and in many instances other materials are added to the polymer to enhance specific properties. In almost all cases this enhancement of a specific property or properties usually results in a corresponding reduction in another property. One of the properties most often affected is that of corrosion resistance. Consequently, care must be taken in specifying thermoplastic polymers. The manufacturer should be checked to verify the specific application.

Table 8.3 provides the allowable operating temperatures for the basic polymer and Table 8.4 lists the atmospheric resistance of the basic polymers. Some additives enhance resistance to atmospheric corrosion, resulting in enhanced protection from UV degradation and increased moisture resistance. Table 8.5 lists additives that are used.

In addition to the polymers whose properties have been modified by additives there are also polymer blends. By blending or alloying two or more polymers a final product can be produced having specific improved properties. Examples of typical blends or alloys are shown below:

ABS-PC (alloy)
PA-phenolic (blend)

By making use of additives, blends, and alloys the resistance to atmospheric corrosion of the basic polymer can be modified either for improvement or for a reduction in its ability to resist atmospheric corrosion. There-

Table 8.3 Allowable Temperature Range of Thermoplastic Polymers[a]

Polymer	Allowable temp. (°F/°C) Minimum	Maximum
ABS	−40/−40	140/60
CPVC	0/−18	180/82
ECTFE	−105/−76	340/171
ETFE	−370/−223	300/149
FEP	−50/−45	400/205
HDPE	−60/−51	180/82
PA	−60/−51	300/149
PAI	<−310/<−190	500/260
PB	—	200/93
PC	−200/−129	250−275/120−135
PCTFE	−105/−76	380/190
PEEK	−85/−65	480/250
PEI	<−310/<−190	500/260
PES	—	340/170
PFA	−310/−190	500/260
PI	<−310/<−190	500−600/260−315
PP	32/0	215/102
PPS	—	450/230
PSF	−150/−101	300/149
PTFE	−20/−29	500/260
PVC	0/−18	140/60
PVDC	0/−18	175/80
PVDF	−50/−45	320/160
UHMWPE	40/4	200/93

[a]Temperature limits may have to be modified depending on the corrodent if in direct contact.

fore, it is essential that the manufacturer be contacted to verify the application.

III. THERMOSETS

Thermosets, or thermosetting polymers, are derived from low molecular weight precursors known as monomors or oligimers. When allowed to polymerize (chemically react) they form continuous network polymers of very

Table 8.4 Atmospheric Resistance of Thermoplastic Polymers

Polymer	UV degradation	Moisture absorption[4]	Weathering	Ozone	SO_2	NO_x	H_2S
ABS	R	0.30	R	X	R[2]		R[2]
CPVC	R	0.03	R	R	R		R[2]
ECTFE	R	<0.1	R	R	R	R	R
ETFE	R	<0.029	R	R	R	R	R
FEP	R	<0.01	R	R	R	R	R
HDPE	RS		R	R	R		R
PA	R	0.6–1.2	R	X	R[1]		X
PC	RS	0.15	L	R			
PCTFE	R		R	R	R	R	R
PEEK	R	0.50	L	R	R		R
PEI	R	0.25	R	R			
PES	RS				X		R
PF					R		R
PFA	R	<0.03	R	R	R	R	R
PI	R		R	R			
PP	RS	0.02	L	L	R	R	R
PPS	R	0.01	R		R		R[2]
PSF	R	0.30	R	R			
PTFE	R	<0.01	R	R	R	R	R
PVC	R	0.05	R	R	R[3]	X	R
PVDC	R		R	R	R	R	R
PVDF	R	<0.04	R	R	R	R	R
UHMWPE	RS	<0.01	R	R	R		R

R = resistant.
RS = resistant only if stabilized with a UV protector.
X = not resistant.
L = limited resistance.
[1]Wet only.
[2]Dry only.
[3]Type 1 only.
[4]Water absorption rate 24 h at 73°F/23°C (%).

Table 8.5 Various Additives and Property Contributions

Additives	Property contribution: Prevent UV degradation	Property contribution: Improve moisture resistance
Alumina powder		X
Carbon black	X	
Coal (powdered)		X
Fibrous glass		X
Kaolin		X
Mica		X
Orlon		X
Silica		X
Talc		X

large molecular weight which are crosslinked to the extent that the network is infusible and insoluble.

Because of the need for strength in structural applications, thermoset polymers are usually used with fiber reinforcement, the most common of which is fiberglass. Since there are now several different materials forming the composite, degradation of the composite can be complex because each material can degrade separately or simultaneously. In addition, there are concerns about the degradation of the interface between the polymer and reinforcement.

As mentioned previously, cure plays an important part in the corrosion resistance of a thermoset polymer, as does the construction of the laminate itself and the type of glass or reinforcing material used. Certain environments may weaken primary and/or secondary polymer linkages, with resulting depolymerization. Other environments may cause swelling or microcracking, whereas still others may hydrolyze ester groupings or linkages. Extraction can occur in certain environments while in other environments repolymerization can take place with a resultant change in structure. Simple solvent action may occur or absorption or attack at the interface between the glass and the resin will result in weakening. The mechanisms involved are complicated and not well understood.

A. Polyesters

The polyesters consist of a series of different unsaturated polyester formulations and are the major thermoset used in reinforced plastics, including those designed for corrosion-resistant applications. Typical corrosion-resistant applications include chemical-resistant piping, chemical storage tanks, floor grating, and structural members such as channel and I beams.

The formulations used primarily for corrosion resistance are:

Bisphenol A–fumarate
Halogenated
Bisphenol A-bisphenol A
Isophthalic

A general-purpose orthophthalate polyester is also produced that is used extensively to manufacture boat hulls.

All formulations are resistant to weathering and UV degradation when stabilized. Although the corrosion resistances of the various formulations differ insofar as handling corrosive chemicals is concerned, their reactions to normal atmospheric corrosion are similar. Their general properties are shown in Table 8.6.

Concentrated sulfuric and nitric acids will attack the polyesters. Although they are resistant to SO_2 and NO_2 gases, adsorption of these gases on wet surfaces can lead to degradation of the polymer.

Operation temperature ranges of the different polyester formulations are given in Table 8.7.

B. Epoxy

Epoxy resins are composed of a large number of varying polymers, hardening agents, fillers, catalysts, and special compounds. Because of the wide variety of epoxy resins, modifiers, and curing agents that are available, the epoxy formulator is able to tailor the epoxy system to meet the application requirement.

At ambient temperatures epoxy formulations can be provided that will resist normal atmospheric corrosion. The general resistance is shown in Table 8.6 and the operating temperature ranges are given in Table 8.7.

C. Vinyl Ester

Vinyl ester resins, like epoxy resins, are composed of a large number of different polymers. It is possible to compound these resins to provide properties to meet the requirement of most applications. There are formulations available that will provide resistance to normal atmospheric pollution as shown in Table 8.6, whereas Table 8.7 gives the operating temperature range of the vinyl esters.

Table 8.6 Atmospheric Resistance of Fiberglass-Reinforced Thermoset Polymers

Polymer	UV degradation	Moisture[a] absorption	Weathering	Ozone	SO_2	NO_2	H_2S
Epoxy	R	0.03	R	R	R	R	R
Polyesters:							
Bisphenol A–fumarate	RS	0.20	R	R	R	R	R
Halogenated	RS	0.20	R	R	R	R	R
Hydrogenated Bisphenol A–bisphenol A	RS	0.20	R	R	R[b]	R	
Isophthalic	RS	0.20	R	R	R	R	R
Teraphthalic	RS	0.20	R	R	R	R	R
Vinyl ester	R	0.20	R	R	R	R	R
Furan	R	2.65	R	R	R	R	R
Phenolic			R				R
Silicone	R	0.02–0.06	R	R	R	R	

[a]Water absorption rate 24 h at 73°F/23°C (%).
[b]SO_3 will cause severe attack.
R, resistant; RS, resistant only if stabilized with UV protector; X, not resistant.

Table 8.7 Operating Temperature Range of Thermoset Polymers[a]

Polymer	Allowable temp. (°F/°C) Minimum	Maximum
Epoxy	−423/−252	300/150
Polyesters:		
Bisphenol A–fumurate		250–300/120–150
Halogenated		250–300/120–150
Hydrogenated bisphenol A–bisphenol A		250–300/120–150
Isophthalic		150/70
Terephthalic		250–300/120–150
Vinyl ester		200–280/98–140
Furan		300–400/150–200
Phenolic		230/110
Silicone	−423/−252	500/260

[a] Temperatures might have to be modified depending on the corrodent.

D. Furans

The series of furan resins are resistant to normal atmospheric pollutants. They weather well and resist UV degradation. Refer to Table 8.6 for the atmospheric resistance of furan and to Table 8.7 for the operating temperature range.

E. Phenolics

Phenolic thermoset resins are the oldest thermoset materials. They do not find as wide an application in the corrosion-resistant field as do the epoxies, vinyl esters, polyesters, and furans. As with other thermoset resins, they can be formulated to meet certain properties, but in general they do not have the range of corrosion resistance of the other thermosets. Table 8.6 shows their resistance to atmospheric corrosion and Table 8.7 gives their operating temperature range.

F. Silicone Resins

Silicone laminates are used as radomes, structures in electronics, heaters, rocket components, ablation shields, coil forms, and terminal boards. Silicone polymers are similar to organic polymers except that certain carbon atoms are replaced by silicon and oxygen atoms. These materials can be used from cryogenic temperatures to 500°F (260°C) as seen in Table 8.7. The silicone polymers exhibit excellent resistance to weathering, UV degradation, ozone, and all normal atmospheric pollutants as seen in Table 8.6.

IV. SUNLIGHT, UV, AND ACCELERATED WEATHERING TESTING

In the discussion of accelerated weathering light sources, the problems of light stability, the effects of moisture and humidity, the effects of cycles, or the reproducibility of results are not taken into account. Simulations of direct sunlight should be compared to what is known as the solar maximum condition: global noon sunlight on the summer solstice at normal incidence. The most severe condition that can be encountered in outdoor service is the solar maximum, which controls the failure of materials. It is misleading to compare light sources against "average optimum sunlight," which is an average of the much less damaging March 21 and September 21 equinox readings.

V. ARC-TYPE LIGHT SOURCE

A. Enclosed Carbon Arc

Since 1918, the enclosed carbon arc has been used as a solar simulator in accelerated weathering and lightfastness testers, and is still specified in many test methods. The UV output of the enclosed carbon arc consists primarily of two large spikes of energy, with very little output below 350 nm. As pointed out previously, the shortest UV wavelengths are the most damaging; consequently, the enclosed carbon arc gives very slow tests on most materials and poor correlations on materials sensitive to short-wavelength UV.

B. Sunshine Carbon Arc (Open Flame Carbon Arc: ASTM G-23)

In 1933 the sunshine carbon arc was introduced. Although it presented an advantage over the enclosed carbon arc, there were still some problems. While the match of sunlight is superior to the enclosed carbon arc, there is still a large spike of energy much greater than sunlight at about 390 nm.

A more serious problem exists in the short wavelengths. The sunshine carbon arc emits a great deal of energy in the UV-C portion of the spectrum, well below the normal solar cutoff point of 295 nm. These short wavelengths can cause unrealistic degradation when compared to natural sources because radiation of this type is found in outer space, not on the earth's surface.

C. Zenon Arc (ASTM G-26)

In 1954 Germany adapted the zenon arc for accelerated weather testing. In order to reduce unwanted radiation, zenon arcs require a combination of filters. Borosilicate inner and outer filters are the most widely used due to borosilicate's giving a close approximation to sunlight by providing a cutoff wavelength of approximately 280 nm, which is relatively close to the sunlight cutoff of 295 nm.

Irradiance settings of 0.35 or 0.55 W/m^2 at 340 mm are the most common. The 0.35 setting compares to winter sunlight whereas the 0.55 setting compares to summer sunlight. However, for practical reasons the 0.35 setting is the most commonly used.

D. Fluorescent UV Lamps

The fluorescent UV testers make use of different lamps with different spectra for different exposure applications. Developed in the 1970s, they use an approach different from that of the arc testers. The fluorescent UV testers do not attempt to reproduce sunlight itself, only the damaging effects of sunlight. Since the short-wavelength UV causes all of the damage to durable materials exposed outdoors, this approach is effective. Consequently, fluorescent UV testers confine their primary light emission to the UV portion of the spectrum.

E. FS-40 Lamp (F40-UVB) (ASTM G-53)

The FS-40 became the first fluorescent UV lamp to be widely used during the early 1970s. The majority of the output of the FS-40 is in the UV-B portion of the UV spectrum, with some output in the UV-A spectrum. Good correlation to outdoor exposures has been achieved with this lamp for the material integrity of plastics. However, the short-wavelength output below the solar cutoff can occasionally cause abnormal results, especially for color retention of plastics.

F. UVB-313 Lamp (ASTM G-53)

The UVB-313 lamp was introduced in 1984. It is essentially a second-generation FS-40 having the same SED, but its output is higher and more stable. As a result of its higher output the UVB-313 provides greater acceleration over the FS-40 for most materials.

G. UVA-340 Lamp (ASTM G-53)

Introduced in 1987, the UVA-340 lamp enhances the correlation of the G-53 devices. The lamp is an excellent simulation of sunlight in the critical short-wavelength UV region, from approximately 365 nm down to the solar cutoff of 295 nm. This lamp shows more realistic testing than the other lamps because the UVA-340 eliminates the short-wavelength output that can cause unnatural results, thereby allowing more realistic testing than many of the other commonly used light sources. The UVA-340 has been tested on plastics and greatly improves the correlation possible with the fluorescent UV and condensation devices.

9
Atmospheric Corrosion of Elastomers

An elastomer is considered to be any material, either natural or synthetic, that is elastic or resilient, and in general resembles natural rubber in feeling and appearance. A more technical definition is provided by ASTM:

> An elastomer is a polymeric material which at room temperature can be stretched to at least twice its original length and upon immediate release of the stress will return quickly to its original length.

Basically an elastomer is a rubber-like material that can be or already is modified to a state exhibiting little plastic flow and quick and nearly complete recovery from an extending force. These materials are sometimes referred to as rubbers. Before modification, an elastomer is referred to as a raw or crude rubber or a basic high polymer.

In order to be considered an elastomer the basic high polymer must be converted, without the addition of plasticizers or other diluents, by appropriate means to an essentially nonplastic state and must meet the following requirements when tested at room temperature (60–90°F) (15–32°C):

1. It is capable of being stretched 100%.
2. After being stretched 100%, held for 5 min, and then released, it can retract to within 10% of its original length within 5 min of release.

Many carbon–hydrogen chain elements tend to gather and behave in basic groups. Hydrogen in a basic group can be substituted by some other element or group to form a new group. These basic groups can react with each other in a head-to-tail fashion to form large molecules and still larger macromolecules. In addition, a group may react with a different group to form a larger group, which can then react with itself in a head-to-tail fashion to produce the new desired macromolecule. The basic groups are known as monomers, and when they are head-to-tail, or polymerize into long chains, they form polymers. Copolymers are the result of the reaction between different basic groups to form a new basic recurring group. These copolymers can also recur in a head-to-tail fashion in long-chain molecules. The basic monomers are usually gases or very light liquids, and as polymerization continues the molecular weight and viscosity both increase until solidification and formation of the gum or solid product results.

Natural rubber is a polymerized hydrocarbon whose commercial synthesis proved to be difficult. Synthetic rubbers that are produced are similar, but not identical, to natural rubber. Natural rubber has the hydrocarbon butadiene as the simplest unit. Butadiene $CH_2{=}CH{-}CH{=}CH_2$ has two unsaturated linkages and is easily polymerized. It is produced commercially by cracking petroleum and also from ethyl alcohol. Natural rubber is a polymer of methyl butadiene (isoprene).

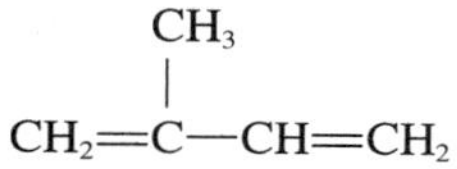

When butadiene or its derivatives become polymerized, the groups link together to form long chains that contain over 1000 groups. In early attempts to produce a synthetic rubber it was found that simple butadiene does not yield a good grade of rubber, apparently because the chains are too smooth and do not interlock with sufficient strength. Better results are obtained by introducing side groups into the chain either by modifying butadiene or by making a copolymer of butadiene and some other compound. As development work continued in the production of synthetic rubber, other compounds were used as the parent material in place of butadiene.

One of the basic monomers is gaseous ethylene, which is polymerized to form polyethylene, which is the basis for other monomers such as styrene, acrylonitrile, isobutylene, vinyl alcohol, and vinyl chloride. These monomers in turn can be polymerized to form polystyrene, polyacrylonitrile, polyisobutylene, polyvinyl alcohol, and polyvinyl chloride. Another basic monomer is butadiene from which the monomers isoprene and chloroprene are derived. These two groups of monomers can react to form copolymers such as butadiene-styrene (GR-S), butadiene-acrylonitrile (Buna-N), and isoprene with isobutylene (butyl), which forms the basis of a basic elastomer series.

Elastomers are primarily composed of large molecules that tend to form spiral threads, similar to a coiled spring, that are attached to each other at infrequent intervals. These coils tend to stretch or compress when a small stress is applied but exert an increasing resistance to the application of additional stresses. This phenomenon is illustrated by the reaction of rubber to the application of additional stress.

In the raw state, elastomers tend to be soft and sticky when hot, and hard and brittle when cold. Compounding increases the utility of rubber and synthetic elastomers. Vulcanization extends the temperature range within which they are flexible and elastic. Table 9.1 provides the operating temperature range of the common elastomers. In addition to vulcanizing agents, ingredients are added to make elastomers stronger, tougher, or harder, to make them age better, to color them, and in general to impart specific properties to meet specific application needs.

I. CAUSES OF FAILURE

Elastomers fail as a result of further chain growth and linkage, which results in a hard, rigid material, or a chain rupture, which results in a plastic or resinous mass. Elastomers in outdoor use are exposed to the deteriorative agents mostly considered in this category, which are sunlight, heat, oxygen, stress with atmospheric ozone, atmospheric moisture, and atmospheric nitrous oxide. Table 9.2 provides the resistance of elastomers to atmospheric pollution. Crosslinkage or chain growth usually decreases elongation and increases hardness and tensile strength, whereas chain rupture will have the opposite effect. Some elastomers will continue to harden and some to soften, whereas others will show an initial hardening followed by softening. All are irreversible reactions.

The low-temperature properties of an elastomer must be taken into account when used in outdoor service. With many elastomers crystallization takes place at low temperatures at which time the elastomer is brittle and will fracture easily. Table 9.3 gives the relative low-temperature flexibility of the more common elastomers. Table 9.4 gives the brittle point of the common elastomers.

II. PROPERTIES OF ELASTOMERS

Although this chapter is dealing primarily with the atmospheric corrosion of elastomers, the other properties of elastomers cannot be ignored when making a selection. It should also be remembered that each of the elastomers can be formulated or blended to improve certain specific properties. This improvement in one specific property may have a detrimental effect on other

Table 9.1 Operating Temperature Range of Common Elastomers

Elastomer	Temp. (°F) Min	Temp. (°F) Max	Temp. (°C) Min	Temp. (°C) Max
• NR natural rubber	−59	175	−50	80
• IR isoprene rubber	−59	175	−50	80
• CR neoprene rubber	−13	203	−25	95
• SBR Buna-S	−66	175	−56	80
• NBR nitrile rubber, Buna-N	−40	250	−40	105
• IIR butyl rubber	−30	300	−34	149
• CIIR chlorobutyl rubber	−30	300	−34	149
• CSM Hypalon	−20	250	−30	150
• BR polybutadiene rubber	−150	200	−101	93
• EA ethylene-acrylic rubber	−40	340	−40	170
• ABR acrylate-butadiene rubber	−40	340	−40	170
• EPDM ethylene-propylene diene monomer	−65	300	−54	149
• SBS styrene-butadiene-styrene		150		65
• SEBS styrene-ethylene-butylene-styrene	−102	220	−75	105
• ST polysulfide	−50	212	−45	100
• FA polysulfide	−30	250	−35	121
• AU polyurethane	−65	250	−54	121
• Polyamides	−40	300	−40	149
• PE polyesters	−40	302	−40	150
• TPE thermoplastic elastomer	−40	277	−40	136
• SI silicone	−60	450	−51	232
• FSI fluorosilicone	−100	375	−73	150
• PVDF polyvinylidene fluoride hexafluoropropene	−40	450	−40	232
• FKM fluoroelastomers	−10	400	−18	204
• ETFE ethylene-tetrafluorethylene	−370	300	−223	150
• ECTFE ethylene chlorotrifluorethylene	−105	340	−76	171
• FPM perfluoroelastomer	−58	600	−50	316

Table 9.2 Resistance of Elastomers to Atmospheric Pollution

Elastomer	UV degradation	Resistance to water absorption	Pollutant				
			Weathering	Ozone	SO_2	NO_x	H_2S
NR natural rubber	P	E	P	P	X		R[2]
IR isoprene	P	E	P	P	X		R[2]
CR neoprene	E	G	E	GE	X	X	R
SBR Buna S	P	GE	P	P			
NBR Buna N	P	FG	F	P	R[1]	X	R[2]
IIR Butyl	G	G	E	E	R		R
CIIR Chlorolbutyl	G	G	F	E			
CSM Hypalon	E	GE	E	E	X	X	R
BR Polybutadiene	P	G	G	P			R[2]
EA Ethylene-acrylic	E	G	G	E			
ABR Acrylate-butadiene	E	G	E	E			
Acrylic ester-acrylic halide	G	G	F	E			
EPDM ethylene-propylene	E	GE	E	E	R	R	R[2]
SBS styrene-butadiene-styrene	P	E	F	P			
SEBS styrene-ethylene-butylene-styrene	G	E	G	E			
ST polysulfide	E	G	E	E			
FA polysulfide	E		E	E			
AU urethene	P	G	G	E			
PA polyamides	P	G	E	E	R		X
PE polyester	G	G	G	E			R
TPE thermoplastic elastomer	G	E	E	E			
SI silicone rubber	E	G	E	E			
FSI fluorosilicone rubber	E	G	E	E			
HFP hexafluoroprene	E	G	E	E			
FKM fluoroelastomers	GE	GE	E	E	X	X	R[1]
ETFE ethylene-tetrafluorethylene	E	G	E	E	R	R	R
ECTFE ethylene-chlorotrifluorethylene	E	E	E	E	R	R	R
FPM perfluoroelastomers	E	G	E	E	R	R	R

E, excellent; G, good; F, fair; P, poor; R, resistant; X, not resistant; 1, wet only; 2, dry only.

Table 9.3 Relative Low-Temperature Flexibility of the Common Elastomers

Elastomer	Relative flexibility
NR natural rubber	G-E
IIR butyl rubber	F
EPDM ethylene-propylene rubber	E
EA ethylene-acrylic	P-F
FKM fluoroelastomer	F-G
FSI fluorosilicone	E
CSM Hypalon (chlorosulfonated polyethylene)	F-G
CR Neoprene (polychloroprene)	F-G
NBR Buna N (nitrile rubber)	F-G
BR polybutadiene	G-E
IR polyisoprene	G
Polysulfide T	F-G
SBR Buna S (butadiene-styrene)	G
SI silicone rubber	E
AU polyurethane	G
EU polyether urethane	G

E, excellent; G, good; F, fair; P, poor.

properties. Consequently, specifications of an elastomer should include the specific properties required for the application, such as those properties described below.

A. Tensile Strength

Tensile strength in the rubber industry refers to the force per unit or original cross-section on elongating to rupture. Since rubbers are rarely used in tensile applications this property in itself is not really important. However, there are other qualities that correlate with tensile strength. As the tensile strength increases, such properties as wear, tear resistance, resilience, cut resistance, stress relaxation, creep, and flex fatigue also improve. When rubber is used as insulation in the electrical industry it should indicate a better cut and abrasion resistance, and generally a tougher and more durable shield.

B. Elongation

Elongation is the maximum extension of an elastomer at the moment of rupture. An elastomer with less than a 100% elongation will usually break if doubled over on itself.

Table 9.4 Brittle Points of Common Elastomers

Elastomer	°F	°C
NR, IR	−68	−56
CR	−40	−40
SBR	−76	−60
NBR	−32 to −40	−1 to −40
CSM	−40 to −80	−40 to −62
BR	−68	−56
EA	−75	−60
EPDM, EPT	−90	−68
SEBS	−58 to −148	−50 to −100
Polysulfide SF	−60	−51
Polysulfide FA	−30	−35
AU	−85 to −100	−65 to −73
Polyamide No. 11	−94	−70
PE	−94	−70
TPE	−67 to −70	−55 to −60
SI	−75	−60
FSI	−75	−60
HFP	−80	−62
FKM	−25 to −75	−32 to −59
ETFE	−150	−101
ECTFE	−105	−70
FPM	−9 to −58	−23 to −50

C. Hardness

The resistance of rubber to deformation is known as hardness and is measured by an instrument known as a durometer, made by the Shore Instrument Company. There are several Shore instruments, namely, A, B, C, D, and O. Each is designed to give different readings from soft sponge up to ebonite-type materials. For elastomeric materials Shore A is the commonly used scale with 0 being soft and 100 hard. By comparison a rubber band would have a reading of approximately 35 and a rubber tire tread 70.

D. Modulus

For elastomeric materials modulus refers to the force per unit of original cross-section to a specific extension, usually 300% elongation, but lower extensions can be used. It is the ratio of the stress of the elastomer to the tensile strain. It differs from that of a metal, as it is not a Young's modulus stress–strain type of curve.

E. Hysteresis

Hysteresis refers to the energy loss per loading cycle. The mechanical loss of energy is converted to heating the elastomer and could reach destructive temperatures.

F. Heat Buildup

The temperature rise in an elastomer resulting from hysteresis is termed heat buildup. It can also refer to use of high frequencies on elastomers where the power factor is high.

G. Permanent Set

Permanent set of an elastomer is the deformation remaining after a given period of stress followed by release of stress.

H. Creep

In elastomers, creep refers to change in strain when the stress is held constant.

I. Compression Set

The permanent creep that remains in an elastomer after it has been held at either constant strain or stress and in compression for a given time is called compression set. Constant strain is most commonly used and is reported as a percentage of the permanent creep divided by the amount of the original strain. A strain of 35% is most common.

J. Abrasion Resistance

Abrasion resistance refers to the resistance of an elastomeric composition to wear.

K. Flex Fatigue

The fracturing of an elastomer after being subjected to fluctuating stresses is known as flex fatigue.

L. Impact Resistance

Impact resistance is the resistance of an elastomer to abrading or cutting when hit by a sharp object.

M. Tear Resistance

Tear resistance is the measure of stress required to continue rupturing an elastomeric sheet, usually after an initiating cut.

N. Flame Resistance

Some elastomers will burn rapidly when ignited whereas others are self-extinguishing when the source of ignition is removed. Flame resistance is the relative flammability of an elastomer.

O. Low-Temperature Properties

The stiffening range and brittle point of an elastomer are indicated by low-temperature properties.

P. Heat Resistance

Time and temperature have an aging effect on all elastomers because none are completely heat-resistant. Heat resistance is measured as change in tensile strength, elongation, and durometer readings from the original values, usually after a 72-h period.

III. SPECIFIC ELASTOMERS

Large quantities of elastomeric materials, both natural and synthetic, are used to produce a myriad of products such as hoses, cable insulation, O rings, seals and gaskets, belting, vibration mounts, flexible couplings, expansion joints, automotive and airplane parts, electrical parts and accessories, and many other items. Many of these applications are exposed to weathering, sunlight, and other forms of atmospheric degradation. Each specific elastomer will be discussed with regard to its applicability to resist atmospheric corrosion. Table 9.5 provides ASTM designation of the elastomers. Many of the elastomers are sold under various tradenames by different manufacturers. Table 9.6 provides a cross-reference listing of generic names, ASTM designation, manufacturers' tradenames, common names, and names of manufacturers providing the different elastomers.

Table 9.5 ASTM Designation of Elastomers

ASTM designation	Elastomer
ABR	Acrylate-butadiene
Butyl IIR	Butyl rubber; GR-1
XNBR	Carboxylic acrylonitrile-butadiene
FPM	Perfluoroelastomers; Chemraz; Kalrez
CIIR	Chlorobutyl rubber
EA	Ethylene-acrylic
ECTFE	Ethylene-chlorotrifluorethylene; Halar
EPDM	Ethylene-propylenediene; Nordel
EPT	Ethylene-propylene terpolymer
FKM	Fluoroelastomers; Viton; Fluorel; Technoflon
IR	Isoprene
NR	Natural rubber
Nitrile NBR	Buna N
BR	Polybutadiene
PE	Polyester; Hytrel
EU	Polyether urethane
AU	Polyurethane
SBR-styrene	Buna S; styrene-butadiene rubber
CR	Polychloroprene; Neoprene
CSM	Chlorosulfonated polyethylene; Hypalon
ACM	Acrylate ester–acrylic halide
SBS	Styrene-butadiene-styrene
SEBS	Styrene-ethylene-butylene-styrene rubber
TPE	Thermoplastic elastomer
SI	Silicone rubber
FSI	Fluorosilicone rubber
HFP	Vinylidene fluoride; PVDF
ETFE	Ethylene-tetrafluorethylene
T	Polysulfide rubber

A. Natural Rubber

Natural rubber (NR) of the best quality is prepared by coagulating the latex of the *Heva brasiliensis* tree, which is cultivated primarily in the Far East. Another rubber is plantation leaf gutta-percha. This material is produced from the leaves of trees grown in bush formation or on plantations. These leaves are plucked and the rubber is boiled out to extract balata, which cures to a hard and tough product. Gutta-percha has been used successfully for more than 40 years as submarine cable.

Chemically, rubber is a polymer of methylbutadiene (isoprene):

$$
\begin{array}{c}
\qquad\quad CH_3 \\
\qquad\quad | \\
CH_2{=}C{-}CH{=}CH_2
\end{array}
$$

When polymerized the units link together to form long chains, each containing over 1000 units.

Purified raw rubber becomes sticky in hot weather and brittle in cold weather. Its valuable properties become apparent after vulcanization.

Depending on the degree of curing, natural rubber is classified as soft, semihard, or hard. Only soft rubber meets the ASTM definition of an elastomer.

During vulcanization, rubber is made to combine with sulfur or sulfur-bearing compounds or with other crosslinking chemical agents. When properly carried out, vulcanization improves mechanical properties, eliminates tackiness, renders the rubber less susceptible to temperature changes, and makes it insoluble in all known solvents. Other materials are added for specific purposes as follows:

1. Carbon blacks, precipitated pigments, and organic vulcanization accelerators are added to increase tensile strength and resistance to abrasion.
2. Whiting, barite, talc, silica, silicates, clays, and fibrous materials are added to cheapen and stiffen.
3. Bituminous substances, coal tar and its products, vegetable and mineral oils, paraffin, petrolatum, petroleum oils, and asphalt are added to soften (for purposes of processing or for final properties).
4. Condensation amines and waxes are added as protective agents against natural aging, sunlight, heat, and flexing.
5. Pigments are added to provide coloration.

Natural rubber has relatively poor weathering and aging properties. When exposed to air in sunlight, rubber tends to become hard and brittle and with time will soften and revert. Its oxygen and ozone resistance is also relatively poor, but compounding can improve this somewhat. Cold water will preserve natural rubber.

Natural rubber finds its major use in the manufacture of pneumatic tires and tubes, power transmission belts, conveyor belts, gaskets, mountings, hose, chemical tank linings, printing press platens, sound and/or shock absorbers, and seals against air, moisture, sound, and dirt.

B. Isoprene Rubber

Isoprene rubber (IR) is the synthetic form of natural rubber, *cis*-polyisoprene. Basically, the properties of isoprene rubber are the same as those of natural

Table 9.6 Elastomer Cross-Reference

Generic name	Designation	Manufacturers,[a] common, or trade name
Natural rubber	NR	26–31
Isoprene	IR	
Polychloroprene	CR	26–31, neoprene (1), Bayprene (2)
Butadiene-styrene	SBR	26–30, Buna-S, GR-S
Butadiene-acrylonitrile	NBR	16, 26–31, nitrile rubber, Buna-N, perbunan (2), Nytek (21)
Butyl rubber	IIR	GR-1, 26–30, Kalar (19)
Chlorobutyl rubber	CIIR	26–30
Carboxylic-acrylonitrile-butadiene	XNBR	16, 26–31
Chlorosulfonated polyethylene	CSM	26–28, 30, 31, Hypalon (1)
Polybutadiene	BR	26–28, 30, 31, Buna-85, Buna-DB (2)
Ethylene-acrylic	EA	13, 28, Vamac (1)
Acrylate-butadiene	ABR	13, 28
Acrylic ester–acrylic halide	ACM	13, 28
Ethylene-propylenediene	EPDM	26–31
Ethylene-propylene terpolymer	EPT	Nordel (1), Royalene, EPDM (8), Dutral (9)
Styrene-butadiene styrene	SBS	Kraton D (3)
Styrene-ethylene-butylene-styrene	SEBS	Kraton G (3)
Polysulfide	ST	27, 28, 30, Thiokol (4)

Polysulfide	FA	Blak-stretchy (14), Blak-tufy (14), Gra-tufy (14)
Urethane	AU	16, 27, 30, 31, 38, Adipene (1) Baytec (2), Futrathane (11), Conathane (16), Texion (2), Urane (23), Pellethane (22), Pure CMC (14)
Polyamides	Nylon	Nylon (1), Rilson (12), Vydyne (18), Plaskon (25)
Polyester	PE	Hytrel (1), Kodar (20)
Thermoplastic elastomers	TPE	Duracryn (1), Flexsorb (17), Geolast (18), Kodapak (20), Santoprene (18), Zurcon (24)
Silicone rubber	SI	27–29, 32, Cohrplastic (15), Green-Sil (14), Parashield (13), Baysilone (2), Blue-Sil (14)
Fluorosilicone	FSI	Parashield (13)
Vinylidene fluoride	HFP	Kynar (7), Foraflon (5)
Fluoroelastomers	FKM	24, 26, 28–31, Viton (1), Fluorel (6), Technoflon (9)
Ethylene-tetrafluorethylene	ETFE	Tefzel (1), Halon-ET (9)
Ethylene-chlorotrifluorethylene	ECTFE	Halar (9)
Perfluoroelastomers	FPM	Kalrez (1), Chemraz (10), Kel-F (6)

[a]List of manufacturers: (1) E. I. Du Pont. (2) Mobay Corp. (3) Shell Chemical Co. (4) Morton Thiokol Co. (5) Atochem Inc. (6) 3-M Corp. (7) Pennwalt Corp. (8) Uniroyal. (9) Ausimont. (10) Green, Tweed, & Co. Inc. (11) Futura Coatings Inc. (12) Attochem Inc. (13) Parker Seal Group. (14) Ther Perma-Flex Mold Co. (15) CHR Industries. (16) Conap Inc. (17) Polymer Corp. (18) Monsanto Co. (19) Hardman Co. (20) Eastman Chemical Products Inc. (21) Edmont Division of Becton, Dickinson & Co. (22) Dow Chemical USA. (23) Krebs Engineers. (24) W. S. Shamban & Co. (25) Allied Signal. (26) General Rubber Co. (27) Hecht Rubber Co. (28) Minor Rubber Co. (29) Newco-Holz Rubber Co. (30) Aldan Rubber Co. (31) Burke Rubber Co. (32) Unaflex.

rubber. The one major difference is that isoprene rubber does not have an odor as does natural rubber.

As far as atmospheric corrosion is concerned, isoprene rubber has no advantages over natural rubber.

C. Neoprene

Chemically, neoprene (CR) is polychloroprene. Its basic unit is a chlorinated butadiene whose formula is

$$\begin{array}{c} \quad\;\; \mathrm{C1} \\ \quad\;\; | \\ \mathrm{CH_2{-}C{-}CH{=}CH_2} \end{array}$$

It is produced from acetylene, which makes this product more expensive than some of the other elastomers.

Neoprene displays excellent resistance to sunlight, weather, and ozone. Because of its low rate of oxidation, products made of neoprene have a high resistance to both outdoor and indoor aging. Neoprene readily accepts color additives. However, lighter shades such as tones of yellow, red, blue, and other bright colors eventually discolor with prolonged exposure to sunlight and ultraviolet (UV) light. Because of this, products intended for outdoor service should be produced in shades of gray, maroon, brown, or black.

Over prolonged periods of time in an outdoor environment, the physical properties of neoprene display insignificant change. When properly compounded, ozone in atmospheric concentrations has little effect on neoprene. When severe ozone concentration is expected, as around electrical equipment, compositions of neoprene can be provided to resist thousands of parts per million of ozone for hours without surface cracking. Natural rubber will crack within minutes when subjected to ozone in concentrations of only 50 ppm.

Neoprene is available in three basic forms:

1. Conventional solid rubber parts
2. Highly compressed cellular materials
3. Free-flowing liquids

Typical solid rubber parts produced are instrument seals, shoe soles and heels, auto spark-plug boots, radiator hose, boating accessories, appliance parts, O rings, and other miscellaneous components.

Cellular forms of neoprene are used primarily for gasketing, insulation, cushioning, and sound and vibration damping.

Fluid neoprene is the primary component in such products as adhesives, coatings and paints, sealants, caulks, and fiber binders.

D. Styrene-Butadiene Rubber (SBR, Buna-S, GR-S)

Japan's occupation of the Far East during World War II created a shortage of natural rubber. Because of the great need for rubber, the U.S. government developed what was originally known as government rubber styrene type. It was later designated GR-S. It has the ASTM designation of SBR and is also known as Buna-S. SBR is produced by copolymerizing butadiene and styrene. The chemical formula is

$$[CH_2—CH{=}CH—CH_2—CH(C_6H_5)—CH_2]$$

In general, Buna-S is very similar to natural rubber. At temperatures below 0°F (−18°C) Buna-S products are more flexible than those produced from natural rubber.

SBR has poor weathering and aging properties. Sunlight will cause it to deteriorate. However, it does have better water resistance than natural rubber.

The major uses of Buna-S is in the manufacture of automobile tires, although some Buna-S materials are also used to manufacture conveyor belts, hose, gaskets, and seals against air, moisture, sound, and dirt.

E. Nitrile Rubber (NBR, Buna-N)

The nitrile rubbers are an outgrowth of the German Buna-N or Perbunan. They are copolymers of butadiene and acrylonitrile:

$$[CH_2{=}C—C{=}N]$$

They are one of the most widely used elastomers. NBR exhibits poor weathering and ozone resistance and deteriorates from exposure to sunlight. It also offers poor general resistance to most atmospheric pollutants, but the nitrile rubbers have excellent resistance to water.

The major applications for Buna-N are in the manufacture of aircraft hose, gasoline and oil hose, and self-sealing fuel tanks. Other applications include carburetor diaphragms, gaskets, cables, printing rolls, and machinery mountings.

F. Butyl Rubber and Chlorobutyl Rubber

Butyl rubber (IIR) contains isobutylene as the parent material to which small proportions of butadiene or isoprene are added. Commercial butyl rubber may contain 5% butadiene as a copolymer. Chlorobutyl rubber (CIIR) is chlorinated isobutylene-isoprene. Both rubbers have the same general properties except that chlorobutyl rubber has a slightly higher allowable operating temperature.

Butyl rubber has excellent resistance to the aging action of air as well as excellent resistance to sun, weather, and ozone. Its weathering qualities are outstanding, as is its resistance to water absorption. Butyl rubber is also resistant to the elements of natural atmospheric pollution.

Butyl rubber, because of its impermeability, finds many uses in the manufacture of inflatable items such as life jackets, life boats, balloons, and inner tubes.

G. Chlorosulfonated Polyethylene (Hypalon)

Chlorosulfonated polyethylene synthetic rubber is manufactured by DuPont under the tradename of Hypalon. In many respects it is similar to neoprene but has better heat and ozone resistance, better electrical properties, and superior weathering resistance. Because of Hypalon's outstanding weathering resistance it is used as an outer protective jacket in high-voltage applications.

Another property that is important in electrical applications is its ability to be colored and not discolor or fade when exposed to sunlight and UV light for long periods. The raw white polymer will accept any color, including pastels, without impairing the true brilliance or hue. Because of Hypalon's natural ozone resistance it is not necessary to add antiozonates during compounding. The antiozonates are strong discoloring agents, which when added to elastomers cause colors to fade and become unstable.

Hypalon is one of the most weather-resistant elastomers available. Oxidation takes place at a very slow rate. Sunlight and UV light have little or no adverse effect on its physical properties. Many elastomers are degraded by ozone concentrations of less than 1 part per million parts of air. Hypalon is unaffected by concentrations as high as 1 part per 100 parts of air.

Hypalon is used to cover automotive ignition and primary wire, nuclear power station cable, control cable, and welding cable. Power and lighting cable on offshore oil platforms are sheathed with Hypalon as an added protection from storms at sea.

Because of Hypalon's color stability and good weathering properties the elastomer is used for exterior parts on trucks, cars, and other commercial vehicles.

Hypalon is also used by the construction industry for sheet roofing, pond liners, reservoir covers, curtain wall gaskets, floor tiles, escalator rails, and decorative maintenance coatings. These applications make use of Hypalon's excellent weatherability, color stability, and good aging characteristics.

H. Polybutadiene Rubber

Butadiene ($CH_2{=}CH{-}CH{=}CH_2$) has two unsaturated linkages and can be polymerized easily. When butadiene or its derivatives become polymerized, the units link together to form long chains each containing over 1000 units.

Polybutadiene, designated BR, is very similar to SBR rubbers but is extremely difficult to process. Consequently, it is used primarily as an admixture with Buna-S and other elastomers.

Polybutadiene has good resistance to water absorption and weathering, but will deteriorate when exposed to sunlight for a long period of time. It also has poor resistance to ozone.

Polybutadiene is very rarely used by itself. It is generally used as a blend with other elastomers to impart better resiliency, abrasion resistance, and/or low-temperature properties, particularly in the manufacture of automobile tire treads, shoe heels and soles, gaskets, seals, and belting.

I. Ethylene-Acrylic Rubber

Ethylene-acrylic (EA) rubbers are produced from ethylene and acrylic acid. The EA rubbers exhibit excellent low-temperature resistance, remaining flexible as low as −50°F (−46°C). They also have extremely good resistance to sun, weather, and ozone. Long-term exposures have no effect on these rubbers. The EA rubbers also have good resistance to water absorption.

Ethylene-acrylic rubber is used in such products as gaskets, hoses, seals, boots, damping components, low-smoke floor tiling, and cable jackets for offshore oil platforms, ships, and building plenum installations. EA elastomers are used in engine parts because of good resistance to heat, fluids, and wear, as well as good low-temperature sealing ability.

J. Acrylate-Butadiene Rubber and Acrylic Ester–Acrylic Halide Rubber

Acrylate-butadiene rubber (ABR) and acrylic ester–acrylic halide rubber (ACM) are very similar to the ethylene-acrylic rubbers having good resistance to sun, weather, and ozone. As a result, the ABR and ACM rubbers find application where resistance to atmospheric conditions and heat are required.

K. Ethylene-Propylene Rubbers

Ethylene-propylene rubber is a synthetic hydrocarbon–based rubber made either from ethylene-propylene diene monomer (EPDM) or ethylene-propylene terpolymer (EPT). These monomers are combined in such a way as to produce an elastomer with a completely saturated backbone and pendant unsaturation for sulfur vulcanization. As a result of this configuration, vulcanizates of EPDM elastomers are extremely resistant to ozone, oxygen, and weather.

Ethylene-propylene rubber is particularly resistant to sun, weather, and ozone attack. Whether the material is formulated in color, white, or black is immaterial since excellent weather resistance is obtained. After years of exposure the elastomer retains a high percentage of its properties and remains free of surface crazing. For all practical purposes EPDM can be considered immune to ozone attack. It is not necessary to add any special compounding ingredients to produce this resistance.

EPDM is used extensively by the automotive industry as a gap-filling panel between the grills and bumper, as well as radiator hose, ignition wire insulation, overflow tubing, window washing tubing, exhaust emission control tubing, body mounts, spring mounting pads, and miscellaneous body seals, floor mats, and pedal pads.

It also finds application in washers being used as door seals and cushions, drain and water circulation hoses, bleach tubing inlet nozzles, boots, seals, gaskets, diaphragms, vibrating isolators, and a variety of grommets. EPDM is also used in dishwashers, refrigerators, ovens, and a variety of small appliances.

The electrical industry also makes wide use of the elastomer in the manufacture of electrical equipment. It is used as an insulating material and in molded terminal covers, plugs, transformer connectors, splices, and insulating and semiconductor tape.

Other manufacturers also take advantage of its weatherability and ozone resistance in the production of such items as garden hose, bicycle tires, sporting goods, tires for garden equipment, dock bumpers, and a variety of other products.

L. Styrene-Butadiene-Styrene Rubber

Styrene-butadiene-styrene (SBS) rubbers are either pure or oil-modified block copolymers and are used primarily in blends with other thermoplastic materials as performance modifiers. They are not resistant to ozone, particularly in the stressed condition, nor are they resistant to prolonged exposure to weather or sun.

M. Styrene-Ethylene-Butylene-Styrene Rubber

The styrene-ethylene-butylene-sytrene (SEBS) rubbers are either pure or oil-modified block copolymer rubbers. They are used as performance modifiers in blends with thermoplastics or as a base rubber for adhesive, sealant, or coating formulation. These rubbers exhibit excellent resistance to ozone. If used for prolonged outdoor exposure applications it is recommended that a UV absorber or carbon black or both be added.

N. Polysulfide Rubbers

Polysulfide (ST and FA) rubbers are also known as Thiokol rubbers. They are manufactured by combining ethylene with an alkaline polysulfide with the sulfur forming a part of the polymerized molecule.

Sodium polysulfide and bis(2-chloroethyl)formal are used to produce ST polysulfide rubber. Products made from these rubbers exhibit outstanding resistance to gas permeation, ozone, and weathering. When compounded with carbon black the ST rubbers are resistant to UV light and sunlight. The resistance of ST rubber to ozone can be improved by the addition of nickel dibutyldithiocarbamate (NBC); however, this addition can degrade the materials' compression set.

The FA polysulfide rubbers are prepared by reacting a mixture of bis(2-chloroethyl)formal and ethylene dichloride with sodium polysulfide. Cured compounds have exceptionally low permeability to gases, water, and organic liquids and exhibit excellent low-temperature flexibility. The FA polysulfide rubber compounds display excellent resistance to ozone, weathering, and exposure to UV light. In general, their resistance is superior to that of the ST polysulfide rubbers. If high concentrations of ozone are to be present, the use of 0.5 part of NBC per 100 parts FA polysulfide rubber will improve the ozone resistance.

Large amounts of these rubbers are used to produce caulking compounds, cements, paint can gaskets, seals, and flexible mountings. Because of the impermeability of the polysulfide rubbers to air and gas, these rubbers have found application in the manufacture of inflatable items such as life jackets, life rafts, and balloons.

O. Urethane Rubbers

The urethane rubbers (UAs) are produced from a variety of polyurethane polymers. The specific polymer used and the compounding determine the properties of the finished product.

In general, the urethane rubbers display excellent resistance to weathering. Extended exposure to UV light reduces their physical properties and causes the rubber to darken. This can be prevented by the use of pigments or UV screening agents.

Urethane rubber products include O rings, electrical encapsulations, gears, tooling pads, small intricate parts, impact mountings, shoe soles, gaskets, bearings, gear couplings, mallets and hammers, solid tires, conveyor belts, and many other items.

P. Polyamides

Polyamides are manufactured under a variety of trade names, the most popular of which is nylon manufactured by DuPont. There are many varieties of polyamides produced, but the four major types are nylon 6, nylon 6/6, nylon 11, and nylon 12. Of these, only nylons 11 and 12 find application as elastomeric materials.

The polyamides have a very low absorption of moisture, which means that parts produced from these materials will have a good dimensional stability regardless of the humidity of the environment. They are resistant to sunlight, weather, and ozone. Polyamide is used to coat many metals to protect them from harsh weather.

Polyamides are used to produce high-quality bicycle seats, shoe soles, gaskets, diaphragms, and seals. Coiled tubing is produced for use as airbrake hose on trucks that have traveled over 2 million miles without a reported failure.

Q. Polyester Elastomer

Polyester (PE) elastomer combines characteristics of thermoplastics and elastomers. They have excellent resistance to ozone and good resistance to general weathering. However, the polyester rubbers do not have satisfactory resistance to sunlight-induced aging unless they are formulated with appropriate additives.

Since the PE polymers do not contain any plasticizers, hoses and tubing produced from them do not stiffen with age. PE is also used to produce seals, gaskets, specialty belting, noise-damping devices, low-pressure tires, industrial solid tires, wire and cable jacketing, electrical connectors, sports equipment, piping clamps and cushions, gears, flexible couplings, and fasteners. Because of their low permeability to air, PE elastomers are also used to manufacture inflatables.

R. Thermoplastic Elastomers

The family of elastomers known as thermoplastic elastomers (TPEs) are based on crosslinked polyolefin alloys and are compounded with common fillers (pigments), plasticizers, stabilizers, and crosslinking agents. The fillers and plasticizers are used to tailor certain properties to specific applications. The TPEs possess excellent weathering resistance and good resistance to sun and ozone. They also have excellent resistance to water.

These elastomers are used in a wide range of applications including reinforced hose, seals, gaskets, flexible tubing, automotive trim, mechanical goods, and wire and cable jacketing.

S. Silicone and Fluorosilicone Rubbers

The silicone (SI) rubbers, also known as polysiloxanes, are a series of compounds whose polymer structure consists of silicone and oxygen atoms rather than the carbon structures of most other elastomers. The silicones are derivatives of silica SiO_2 or O═Si═O. Silicone rubber is produced by breaking the double bonds and introducing methyl groups into the linkages as shown below:

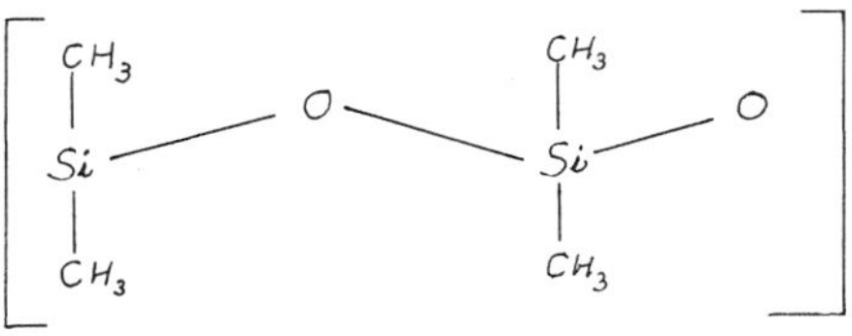

Silicone, although in the same chemical group as carbon, is a more stable element than carbon; therefore more stable compounds can be produced. The basic structure may be modified with vinyl or fluoride groups.

The silicone and fluorosilicone (FSI) rubbers exhibit excellent resistance to sun, weathering, ozone, and water. Their excellent weathering qualities and wide temperature range have led to their widespread use in caulking compounds.

T. Vinylidene Fluoride Rubbers

Polyvinylidene fluoride (PVDF) is a homopolymer of 1,1-difluoroethane with alternating CH_2 and CF_2 groups along the polymer chain. When exposed to aggressive chemicals this polymer has the characteristic stability of fluoropolymers. Hexafluoropropene (HFP) is highly resistant to sunlight, weather, and ozone. Because of its resistance to weathering, UV light, radiation, and fungi, it finds many applications that make use of these characteristics. In the electrical and electronics field, PVDF is used for multiwiring jacketing, plenum cables, heat-shrinkable tubing, computer wiring, and cable ties.

U. Fluoroelastomers

The fluoroelastomers (FKMs) are manufactured under various tradenames by different manufacturers. Typical materials are:

Tradename	Manufacturer
Viton	DuPont
Technoflon	Ausimont
Fluorel	3M

FKM elastomers are fluorine-containing hydrocarbon polymers with a saturated structure obtained by polymerizing fluorinated monomers such as vinylidene fluoride, hexafluoropropene, and tetrafluoroethylene. In addition to being able to operate at elevated temperatures they are also generally serviceable in dynamic applications down to −10°F (−23°C). Thickness of material determines the flexibility at low temperatures, i.e., the thinner the material the more flexible. The brittle point at a thickness of 0.075 in. (1.9 mm) is in the neighborhood of −50°F (−45°C). Depending on the thickness and hardness of the material this temperature can have a range of −25 to −75°F (−32 to −59°C).

The fluoroelastomers are relatively impermeable to air and gases, ranking about midway between the best and poorest elastomers in this respect.

As a result of their chemically saturated structure, the fluoroelastomers display excellent weathering resistance to sunlight and especially to ozone. Samples exposed to various tropical conditions in Panama for 10 years showed little or no change in properties. Similar results were experienced after 13 years of exposure in direct Florida sunlight.

FKM elastomers will resist ozone concentrations as high as 100 ppm. This is an important property when it is recognized that standard tests require resistance to only 0.5 ppm ozone.

Because of their high cost, the fluoroelastomers are mainly used in applications with severe operating conditions that no other elastomer can meet. Many applications are found in the aerospace industry as a result of their reliability under extreme exposure conditions.

V. Ethylene-tetrafluorethylene Elastomer

Ethylene-tetrafluorethylene elastomer (ETFE) is a modified partially fluorinated copolymer of ethylene and polytetrafluorethylene (PTFE). ETFE elastomers have outstanding resistance to sunlight, ozone, and weather. This feature, coupled with its wide range of corrosion resistance, makes the material particularly suitable for outdoor applications subject to atmospheric corrosion.

W. Ethylene-chlorotrifluorethylene Elastomer

Ethylene-chlorotrifluorethylene elastomer (ECTFE) is a 1:1 alternating copolymer of ethylene and chlorotrifluorethylene. ECTFE is extremely resistant to sun, weather, and ozone attack. It also has a broad-use temperature range, from cryogenic to 340°F (171°C). Because of these properties it is suitable for application subject to outdoor atmospheric corrosion.

X. Perfluoroelastomers

The perfluoroelastomers (FPMs) are true rubbers, as compared to other elastomeric compounds. They are more resistant to swelling and embrittlement and retain their elastomeric properties over the long term. These elastomers display excellent resistance to sunlight, weather, and ozone.

As with the fluoroelastomers, applications of the FPM elastomers is limited to those areas where other less expensive elastomers will not perform satisfactorily. These are areas of extreme temperatures and highly corrosive conditions. They are suitable for outdoor applications to resist atmospheric corrosion.

IV. SUMMARY

When selecting an elastomer there are many properties that must be taken into account besides its resistance to atmospheric corrosion. One factor that must be considered is the ability of the elastomers to be compounded. By blending various additives, specific elastomer properties may be modified. Invariably while compounding improves one specific property, it is at the expense of another property. For this reason it is important that the manufacturer be consulted and made aware of the application and the specific specifications to be met.

It should also be remembered that ambient temperature is not always 70°F (23°C). Outdoor applications, depending on the geographic location, may be subject to wide temperature fluctuations during different seasons of the year. Although most elastomers are capable of operating at an upper ambient temperature range, many are not capable of operating at some of the potential lower temperature ranges.

REFERENCE

1. PA Schweitzer. Corrosion Resistance of Elastomers. New York: Marcel Dekker, 1990.

10
Atmospheric Corrosion of Masonry Materials

Concrete and cement-based products are widely used construction materials. The primary ingredient is Portland cement, which is also used to produce common building block. Portland cement was patented in 1824 by an English bricklayer named Joseph Aspin. He produced the first batch by burning limestone and clay together. The name Portland cement was derived after a natural stone found on the Isle of Portland.

Portland cement is made from limestone or other natural sources of calcium carbonate, clay (a source of silica), alumina, ferric oxide, and minor impurities. After grinding, the mixture is fired in a kiln at approximately 2500°F (1371°C). The final product is ground to a fineness of about 10 μm and mixed with gypsum to control setting. In order to produce a concrete mix of the desired consistency and mechanical properties, it is combined with sand, stone, and water.

As the concrete hardens chemical reactions take place. The two most important reactions are the generation of calcium hydroxide and of tricalcium silicate hydrate. The calcium hydroxide generated could theoretically be as high as 20% of the weight of the cement, producing an alkalinity that at the solubility of lime results in an equilibrium pH of 12.5.

Tricalcium silicate hydrate (tobermorite gel) is the compound most responsible for the mechanical strength and dimensional stability of the hardened concrete. The pore space within the hardened concrete leading to its adsorption, porosity, and drying shrinkage is the result of the tobermorite gel structure.

Tobermorite gel particles have an average particle size of approximately 1000 times that of the original unhydrate cement grain of 10 μm. This large increase in surface area produces the forces responsible for the cementing action of the product.

The volume of pore spaces is also the result of the tobermorite gel. In addition to these pores there are voids caused by capillary movement of excess water and air entrapped by the mixing operation. The resulting concrete can have a water adsorption capacity in excess of 20% of the apparent volume of the mass.

The total amount of water used in the mix influences the volume of pore spaces. In order to complete the hydration of the cement, water equal to approximately 25% of the weight of the cement is required. However, this amount is not sufficient to permit placing of the concrete mix. To make the mix fluid, an additional 25% is added making a total water/cement ratio of 0.5. After hardening, this excess water evaporates out of the mix, contributing to void formation. These voids are important because the porosity is responsible for the adsorption of water into the bulk mass.

In cold climates the internal water may freeze, causing concrete spalling. The use of deicing salts can also cause spalling. The salt solution penetrates the concrete surface and after the water evaporates, the salts crystallize within the void spaces, creating internal pressure that causes the spalling.

I. ATMOSPHERIC ATTACK OF CONCRETE

Although concrete and all cement-based products are inherently weak in tension, they are strong in compression. To overcome the weakness in tension, steel rods (reinforcing rods) are placed in the uncured mix. The reinforcing steel can be plain or pre- and/or poststressed. Stressed steel places the concrete in compression (its strong point). Any tensile load placed on the structure must overcome the compressive load due to the stressed steel before the concrete will be placed in the tensile load mode. Micro- and macrocracking of the concrete results due to the weakness of tensile loading ability, which reduces the life expectancy in a corrosive environment. Corrosives gain access to the interior of the concrete through these cracks. If this results in the rusting of the embedded steel, then the volume of the excess iron oxide cannot be accommodated. Because of the poor tensile strength of the concrete, spalling of the cement mass will take place.

The resistance of concrete to corrosion is the result of its nonoxidizable structure (resistance to water and oxygen). Steel that has been embedded in the mix is passivated as a result of the hardened material's 12.5 pH. Concrete will resist degradation as long as nothing in the environment dissolves the cement matrix or reduces its ability to passivate the embedded steel.

The alkalinity of the concrete is provided by the presence of calcium oxide (lime). Any material that will cause the calcium oxide or hydroxide to be removed, lowering the pH, will prove detrimental and cause solution of the cement hydrates. Contact with inorganic or organic acids can cause this to happen. Organic acids can be generated when organic materials ferment.

When carbon dioxide dissolves in water that may be present on the concrete, weak carbonic acid is produced. The weak carbonic acid lowers the pH of the cement solution, allowing the embedded steel to corrode. This is sometimes referred to as the carbonation of cement.

Sulfates will also cause concrete to deteriorate. In addition to being able to produce sulfuric acid, which is highly corrosive to concrete, sulfates are also reactive with some aggregates used in concrete mixes. Sulfate ions react with tricalcium aluminate to form sulfoaluminate hydrate with a large crystallized water content.

II. PROTECTING CONCRETE

Several steps can be taken to prolong the life of concrete. The first is to protect the reinforcing steel from corrosion by applying a rust preventative to the surface prior to embedding in the concrete mix. This keeps the steel from rusting if contacted by water and causing spalling of the concrete.

A second approach is to provide a coating for the concrete. Before selecting an appropriate coating, consideration must be given to the condition of the concrete and the environment to which the concrete will be exposed. Proper surface preparation is essential. Surface preparations can be different for freshly placed concrete than for old concrete.

When concrete is poured it is usually held in place by means of steel or wood forms that are removed when the concrete is still in its tender state. To facilitate their removal release agents are applied to the forms prior to pouring. Oils, greases, and proprietary product release agents are left on the surface of the poured concrete. They must be removed if they will interfere with the adhesion of subsequent coatings.

Quite often curing compounds are applied to fresh concrete as soon as practical after the forms have been removed. These are liquid membranes based on waxes, resins, chlorinated rubber, or other film formers, usually in a solvent. Pretesting is necessary to determine whether or not they will interfere with the coating to be applied.

Generally admixtures that are added to concrete mixes in order to speed up or slow down the cure, add air to the mix, or obtain other special effects will not interfere with surface treatments to improve durability. The concrete supplier can furnish specific data regarding the admixtures. If in doubt, try a test patch of the coating material to be used.

It is essential that water from subslab ground sources be eliminated or minimized since migration through the concrete may create pressures at the bond line of water-resistant barriers.

A. Surface Preparation

It is essential that the concrete surface be properly prepared prior to application of the coating. The surfaces of cement-containing materials may contain defects that require repair prior to the application of the coating. In general, the surface must be thoroughly cleaned and all cracks repaired.

Unlike specifications for the preparation of steel prior to coating, there are no detailed standard specifications for the preparation of concrete surfaces. In most instances it is necessary to follow the instructions supplied by the coating manufacturer. Specifications can range from simple surface cleaning that provides a clean surface without removing concrete from the substrate, to surface abrading that provides a clean roughened surface, to acid etching that also provides a clean roughened surface.

B. Surface Cleaning

Surface cleaning is accomplished by one of the following means:

1. Broom sweeping
2. Vacuum cleaning
3. Air blast
4. Water cleaning
5. Detergent cleaning
6. Steam cleaning

Water cleaning and detergent cleaning will not remove deep embedded soils in the pores of the concrete surfaces. In addition, these methods saturate the concrete, which then requires a drying period (not always practical).

C. Surface Abrading

Surface abrading can be accomplished by:

1. Mechanical abrading
2. Water blasting
3. Abrasive blasting

Of the three methods to produce a roughened surface, the most technically effective is abrasive blasting by means of sandblasting or shot blasting followed by broom sweeping, vacuum cleaning, or an air blast to remove the abrasive. However, in some instances this may not be a practical approach.

D. Acid Etching

Acid etching is a popular procedure used for both new and aged concrete. It must be remembered that during this process acid fumes will be evolved that may be objectionable. Also, thorough water rinsing is required, which saturates the concrete. This may necessitate a long drying period, depending on the coating to be used.

III. COATING SELECTION

The coating to be selected will depend on the atmospheric conditions to be encountered. Not only will the normal atmospheric pollutants have to be considered but any specific local pollutants will also have to be taken into account. Also to be considered are the local weather condition, which will result in minimum and maximum temperatures as the seasons change. There are many monolithic surfacings or coatings from which to choose. Any selection must be based on the resistance required, life expectancy, maintenance costs, and initial costs including necessary surface preparation.

The most popular monolithic surfacings are formulated from the following resins:

1. Epoxy including epoxy novolac
2. Polyester
3. Vinyl ester including vinyl ester novolac
4. Acrylic
5. Urethane

Monolithic surfacings are installed at thicknesses at 1/16 in. (1.5 mm) to 1/2 in. (13 mm).

A. Installation of Monolithic Surfaces

As with the application of any coating system, proper surface preparation and installation techniques are essential. Basic preparations for the application of a coating are as follows:

1. Substrate must be free of cracks and properly sloped to drains.

2. New as well as existing slabs should have a coarse surface profile, be clean and dry, and be free of contaminants.
3. In general, slab and materials to be installed should have a temperature of 65–85°F (18–29°C). If necessary, special catalysts and hardening systems are available to accommodate higher or lower temperatures.
4. Manufacturer's directions should be adhered to regarding the priming of the substrate prior to applying the monolithic surfacing.
5. Individual and combined components should be thoroughly mixed at a maximum speed of 500 rpm to minimize air entrainment.
6. Prior to curing, the surface must be protected from moisture and contamination.

Monolithic surfacings may be installed by a variety of methods, many of which are the same methods used in the Portland cement concrete industry. The primary methods used are as follows:

1. Hand-troweled
2. Power-troweled
3. Spray
4. Pour-in-place/self-level
5. Broadcast

Small areas, or areas with multiple obstructions, should be hand-troweled. Topcoat sealers should be applied to provide imperviousness and increased density, with a smooth, easy-to-clean finish.

Large areas with a minimum of obstructions are best handled with power troweling. The minimum thickness would be 1/4 in. (6.5 mm). The density of the finish can be improved by the use of appropriate sealers.

For areas subjected to aggressive corrosion, spray application is recommended. The consistency of the material can be formulated to control slump and type of finish. This permits the material to be sprayed on vertical and overhead surfaces including structural components, with thicknesses of 1/16–3/32 in. (1.5–2.4 mm) as is suitable for light-duty areas.

Economical and aesthetically attractive floors can be applied by the broadcast system, in which resins are "squeegee"-applied to the concrete slab. Filler or colored quartz aggregates of varying color and size are sprinkled or broadcast into the resin. Excess filler and quartz are vacuumed or swept from the floor after the resin has set. This results in a floor thickness of 3/32–1/8 in. (2–3 mm). This type of floor is outstanding for light industrial and interior floors.

B. Epoxy and Epoxy Novolac Surfacings

The two most often used epoxy resins for monolithic surfacings are the bisphenol A and bisphenol F (epoxy novolac). Epichlorohydrin is reacted with the base component to form resins of varying viscosity and molecular weight. The hardening system employed determines the molecular orientation. Properties dictated by the hardening systems are as follows:

1. Chemical and thermal resistance
2. Physical properties
3. Moisture tolerance
4. Workability
5. Safety during use

Aliphatic amines, modified aliphatic amines, and aromatic amines are the primary hardeners used. Of these the aromatic amines provide the widest range of corrosion resistance. Amine hardening systems, being alkaline, are extremely compatible for application to poured-in-place and precast concrete, and are the most popular for ambient temperature curing of monolithic surfacings. Since these systems are hygroscopic and can present allergenic responses to sensitive skin, protective garments including gloves should be utilized when employing epoxy materials.

Epoxies are economical, available in a wide range of formulations and properties, and offered from many manufacturers. Formulations are available for interior as well as exterior applications. However, when installed outside, moisture from beneath the slab can affect adhesion and cause blistering.

The typical epoxy system is installed in layers of primer, base, and finish coats. Overall, the installation can take several days. Epoxies should not be applied to new concrete before it has reached full strength (approximately 28 days).

Table 10.1 provides the atmospheric corrosion resistance of monolithic floor surfacings.

The bisphenol F's (epoxy novolacs) are essentially premium grade epoxy resins providing an increased chemical resistance. They are also available in a wide range of formulations and from many manufacturers. The primary advantages in the use of epoxy novolacs is in their improved resistance to higher concentrations of oxidizing and nonoxidizing acids, and aliphatic and aromatic solvents.

The novolacs are more expensive than the bisphenol A epoxies and can discolor from contact with sulfuric and nitric acids. For exposure to normal atmospheric pollutants the bisphenol A epoxies are satisfactory. However, if the surface is to be exposed to other more aggressive contaminants, then the novolacs should be considered.

Table 10.1 Atmospheric Corrosion Resistance of Monolithic Concrete Surfacings

Surfacing	Atmospheric pollutant							
	NO_x	H_2S	SO_2	CO_2	UV	Chloride salt	Weather	Ozone
Epoxy–bisphenol A								
Aromatic amine hardener	R	X	X	R	R	R	R	R
Epoxy novolac	R	X	X	R	R	R	R	R
Polyesters:								
Isophthalic	R	R	X	R	RS	R	R	R
Chlorendic	R	R	X	R	RS	R	R	R
Bisphenol A fumurate	R	R	X	R	RS	R	R	R
Vinyl esters	R	R	R	R	R	R	R	R
Acrylics	R		R	R	R	R	R	R
Urethanes	R		X		R	R		

R, resistant; X, not resistant; RS, resistant when stabilized.

C. Polyester

Unsaturated polyester resins are used for monolithic surfacings. The three types that are primarily used are as follows:

1. Isophthalic
2. Chlorendic acid
3. Bisphenol A fumarate

The chlorendic and bisphenol A fumarate resins offer improved chemical resistance over the isophthalics (see Table 10.1). The essentially neutral curing systems provide capability for application to concrete substrates. When properly formulated the polyester resins provide installation flexibility to a wide range of temperatures, humidity, and contaminants.

D. Vinyl Ester and Vinyl Ester Novolac

Vinyl ester resins are the most corrosion-resistant of any of the monolithic surfacings, and they are also the most expensive and difficult to install. They are used when extremely corrosive conditions are present. The finished flooring is vulnerable to hydrostatic pressure and vapor moisture transmission. Refer to Table 10.1 for their corrosion-resistant properties.

E. Acrylic

Acrylic monolithic coatings are suitable for interior or exterior applications with relatively benign atmospheric exposures. They excel at water and weather resistance, and are best at "breathing" in the presence of a moisture transmission problem in the slab. However, their overall corrosion resistance is limited (see Table 10.1).

F. Urethane

The urethane systems are similar to the acrylics in that they are both applicable for use in areas subject to light corrosion. They can be used in temperature ranges of −10 to 140°F (−24 to 60°C).

The urethanes will retain their appearance (gloss retention, nonchalking, and fading) for extended periods of time. Care must be taken during storage and application of the urethanes because they are moisture-sensitive prior to full cure. Contamination will cause foaming, blistering, and loss of the glossy appearance of a coated substrate (see Table 10.1).

IV. POLYMER-MODIFIED PORTLAND CEMENT CONCRETE

Polymer-modified Portland cement concrete is normal concrete to which a synthetic resin has been added in the formulation. The same generic resins can be used that are used in polymer concretes. Properties of these two concretes differ, even when the same generic resin is used in both.

While the polymer concrete may be totally chemical-resistant the comparable polymer modified Portland cement concrete is not, although certain benefits are derived including:

1. Reduction in the absorption of the concrete
2. Improved impact resistance
3. Allowable application of concrete in thinner cross-sections
4. Improved adhesion for pours onto existing concrete
5. Improved resistance to salt, but not to aggressive corrosive chemicals

V. POLYMER CONCRETE

Polymer concrete is a chemically resistant synthetic concrete consisting of low-viscosity binders and inert aggregates, which when combined and thoroughly mixed produces a synthetic material that can be precast or poured in place. They may also be used as a concrete surfacing. When applied in thicknesses greater than 1/2 in. (13 mm) it is considered a polymer concrete. When applied in lesser thickness the material is referred to as a surfacing or coating.

Polymer concretes are not the same as polymer modified Portland cement concrete. Polymer concretes are composed totally of synthetic resin compounds with fillers. They do not contain any Portland cement. Polymer-modified Portland cement concrete can use some of the same generic resins as used in polymer concretes, but with different results.

The primary resins used to formulate polymer concretes are as follows:

1. Furan
2. Epoxy, including novolac
3. Polyester
4. Vinyl ester, including vinyl ester novolac
5. Acrylics
6. Sulfur

Before pouring polymer concrete mixes, a liquid-tight membrane should be placed over the substrate. Such a membrane will act as a liquid barrier and separate the polymer concrete from the surface below it, allowing it to move as a unit, sliding on the membrane. This permits movement during curing, temperature changes, etc. The substrate must be smooth and penetrations through the polymer concrete be centered between joints.

Depending on the service, polymer concrete is cast in thicknesses from 1/2 in. (13 mm) to several inches, either with or without reinforcing. Castable polymer concretes are mixed and cast in a manner similar to that of Portland cement concrete, being poured into forms to harden. Troweled applications are generally thin coatings of less than 3/4 in. (20 mm) and are troweled in place with or without reinforcement.

Like all cementatious materials, polymer concretes are inflexible and tend to brittleness. The modulus of elasticity range from 10^5 to 10^6 psi and flexural strength from 500 to 1000 psi. Tensile strengths vary from 150 to 400 psi and compressive strengths range from 1800 to 5000 psi.

A. Furan Polymer Concrete

Most synthetic resins are petrochemical-based. Furan resins are unique in that they are agriculturally based, with the furfurayl alochol produced from agricultural byproducts such as corn cobs, bagasse, rice, and oat hulls. Although the most versatile of all resins in use, furan resins are not a popular choice for use as polymer concrete because of their tendency to shrink and their brittleness. They do provide a wide range of corrosion resistance and are resistant to atmospheric corrosion (see Table 10.2).

B. Epoxy Polymer Concrete

The same epoxy resins used to formulate monolithic surfacings are also employed to produce polymer concrete. As indicated previously, the resins of choice are as follows:

1. Bisphenol A
2. Bisphenol F (epoxy novolac)

For overall corrosion resistance the epoxy novolacs are somewhat superior to bisphenol A. However, for strictly atmospheric corrosion resistance bisphenol A is satisfactory. The filler most often used with epoxy concretes is silica along with an amine hardening system. This combination provides a polymer concrete with outstanding physical properties and moderate overall corrosion resistance. However, it is perfectly suitable to resist normal atmospheric corrosion (see Table 10.2).

Table 10.2 Resistance of Polymer Concretes to Atmospheric Corrosion

Polymer concrete	Atmospheric pollutant							
	NO_x	H_2S	SO_2	CO_2	UV	Chlorides	Weather	Ozone
Furan	X	R	R	R		R	R	
Epoxy–bisphenol A with aromatic amine hardener	R	X	X		R	R	R	R
Epoxy novalac	R	X	X		R	R	R	R
Polyester resins:								
Chlorendic type	R	R	X	R	RS	R	R	R
Bisphenol A fumurate type	R	R	X	R	RS	R	R	R
Acrylics	R		R	R	R	R	R	R
Sulfur	R	X	R	R	R	R		

R, resistant; RS, resistant when stabilized; X, not resistant.

C. Polyester Polymer Concrete

Unsaturated polyester resins formulated with either carbon or silica fillers are used as polymer concretes. Bisphenol A fumarate resins and chlorendic acid resins are the two primarily used to formulate polyester concrete. Either resin can provide corrosion protection from normal atmospheric pollutants. Of the two, the chlorendic resins have a somewhat superior overall corrosion resistance (see Table 10.2).

D. Vinyl Ester Polymer Concrete

Vinyl ester resins are products resulting from the addition reactions of methacrylic acid and epoxy resin. They possess many of the properties of epoxy, acrylic, and polyester resins, which results in their exceptional corrosion resistance and physical properties.

The vinyl ester resins have lower shrinkage and are generally less rigid than many of the polyester resins. Vinyl ester novolac is a premium vinyl ester resin. Any of the vinyl ester resins can be formulated to produce a polymer concrete that will be capable of resisting normal atmospheric pollution.

Either the vinyl ester or vinyl ester novolacs are suitable for use where normal atmospheric corrosion is present. Refer to Table 10.2 for the corrosion resistance of the vinyl esters.

E. Acrylic Polymer Concrete

The acrylic resins possess several advantages:

1. They are the easiest of the resin systems to mix and apply by using poured-in-place and self-leveling techniques.
2. They exhibit outstanding weather resistance.
3. The acrylic formulations are the only ones that can be installed at below freezing temperatures, 25°F (−4°C).
4. They have the fastest set and cure time of any of the resins. They develop 90% of their ultimate strength in 4 h.
5. They are the easiest to pigment.

Acrylic polymer concrete is particularly suitable in areas subject to normal atmospheric corrosion. It exhibits good weather resistance, as well as resistance to airborne SO_2, SO_3, and NO_x. Refer to Table 10.2 for additional information.

F. Sulfur Polymer Concrete

Sulfur polymer concrete predates organic resin counterparts. It is strictly an acid-resistant formulation limited to a maximum pH 7. Fillers of 100% silica or 100% carbon are used.

Because of its flammability and the potential for oxides of sulfur, it is not normally used for most indoor applications. Table 10.2 provides the resistance of sulfur concrete to atmospheric corrosion.

VI. MORTARS AND GROUTS

Chemically resistant mortars and grouts are based on organic and inorganic chemistry. They consist of a liquid resin or an inorganic binder; fillers such as silica, carbon, or combinations of the same; and a hardener or catalyst system that may be incorporated in the filler or added as a separate component.

A mortar is a material having a heavy consistency that can support the weight of a brick or tile without being squeezed from between the joints while curing.

A grout is a thin or "soupy" mortar used for filling joints between previously laid brick or tile. Chemically resistant machinery grouts contain larger aggregates than tile grout and have a different resin viscosity.

The effectiveness and workability of a mortar or grout depends on the proper selection of fillers, or combination of fillers, particle size, resin viscosity, and reactivity of catalysts and hardeners. Hardener systems must be formulated for application at temperature ranges of approximately 60–90°F (15–32°C). The lower the temperature the slower the set; the higher the temperature the faster the set.

Carbon, silica, or combinations of the two are the most popular fillers used. Carbon is the most inert of the fillers and therefore is the one most often selected. The most popular resin systems, both organic and inorganic, will be discussed, starting with the organic systems, many of which are also used as monolithic surfacings.

A. Epoxy

The same basic epoxy resin systems are used as monolithic surfacings and epoxy polymer concrete. In addition to these, epoxy phenol novolac is also used to produce mortars and grouts. This is a higher viscosity resin that requires the inclusion of various types of diluents and resin blends. Table 10.3 lists the atmospheric corrosion resistance of the mortars and grouts.

Table 10.3 Atmospheric Corrosion Resistance of Mortars and Grouts

	Atmospheric pollutant							
Mortar/grout	NO_x	H_2S	SO_2	CO_2	UV	Chloride salt	Weather	Ozone
Epoxy with aromatic amine hardener	R	X	X	R	R	R	R	R
Furans	X	R	R	R		R	R	
Phenolics	X	R	R	R		R		
Polyesters	R	R	X	R	RS	R	R	R
Vinyl esters	R	R	R	R	R	R	R	R
Sulfur	R	X	R	R	R	R	R	
Sodium silicate:								
Normal	X	R	X		R	R		
Water-resistant	R	R	R		R	R	R	
Potassium silicate with setting agent:								
Organic	X	X	X	R	R	R		
Inorganic	R	R	R	R	R	R	R	
Silica	R	R	R	R	R	R	R	

R, resistant; RS, resistant when stabilized; X, not resistant.

B. Furans

Furan mortars and grouts are formulated using either 100% carbon or part carbon/silica fillers. Although not normally used as a monolithic surfacing, they are used to produce furan polymer concrete. Either formulation is satisfactory to resist normal atmospheric pollution but more aggressive pollutants might require the use of 100% carbon filler. Furan grouts normally are filled with only 100% carbon filler.

The furan resin and grouts are two-component systems consisting of resin and filler. The filler contains the catalyst or hardener.

Refer to Table 10.3 for atmospheric corrosion resistance.

C. Phenolics

Phenolic resins are not used in monolithic surfacings in grouts or in polymer concretes but are extensively used in mortars. They are two-component systems as are the furan resins using 100% carbon filler, or 100% silica filler, or part carbon and part silica filler. Like the epoxies, the phenolic resins can be allergenic to sensitive skin. Protective clothing should be worn by persons handling the phenolic resins. Phenolic resins have a limited shelf life and must be stored at 45°F (7°C).

For resistance to normal atmospheric corrosion any of the fillers are suitable. Refer to Table 10.3 for the atmospheric corrosion resistance.

D. Polyesters

The same polyester resins are used to formulate mortars and grouts as are used to formulate monolithic surfacings and polymer concretes. Carbon and silica fillers are used in the formulations. Polyester resins have a shelf life limitation and should be stored below 60°F (15°C). Any of the polyester-type resins provide suitable resistance to normal atmospheric corrosion (see Table 10.3).

E. Vinyl Esters

The same vinyl ester formulations are used for mortars and grouts as are used for monolithic surfacings. They have a limited shelf life and should be stored below 60°F (7°C). Any of the formulations provide suitable protection from normal atmospheric corrosion (see Table 10.3).

F. Sulfur

Sulfur mortars and grouts are similar to sulfur polymer concrete. Fillers used are either 100% carbon or 100% silica. The sulfur mortars and grouts are easy to use, have high early strength and unlimited shelf life, and are economical. They exhibit good resistance to normal atmospheric corrosion (see Table 10.3).

G. Sodium Silicates

The sodium silicates are available as two-component systems consisting of filler and binder. The filler contains the setting agent. These silicates are referred to as soluble silicates because they are soluble in water. Although resistant to strong acids they cannot be used in contact with weak acids.

New setting agents have been developed that improve the water resistance of the sodium silicate mortars making them suitable for use with dilute as well as concentrated acids (see Table 10.3 for their resistance to normal atmospheric corrosion).

H. Potassium Silicates

Potassium silicate mortars are available as two-component systems, filler and binder, with the setting agent in the filler. The properties of the mortar are determined by the setting agent, which can be organic or inorganic. Organic setting agents are water-soluble and can be leached out if the mortar is exposed to steam or moisture. Mortars produced with inorganic setting agents are water- and moisture-resistant. Refer to Table 10.3 for their resistance to normal atmospheric corrosion.

I. Silica (Silica Sol)

Silica (silica sol) mortars consist of a colloidal silica binder with quartz fillers. They are more difficult to use than the other silicate mortars. Refer to Table 10.3 for their resistance to normal atmospheric corrosion.

REFERENCES

1. WG Read, Jr, and CE Zimmer. Cements and mortars. In: PA Schweitzer, ed. Corrosion and Corrosion Protection Handbook, 2nd ed. New York: Marcel Dekker, 1988, pp. 521–531.
2. AA Boova. Chemical-resistant mortars, grouts, and monolithic surfacings. In: PA Schweitzer, ed. Corrosion Engineering Handbook. New York: Marcel Dekker, 1996, pp. 459–487.
3. WL Sheppard, Jr. Chemically Resistant Masonry, 2nd ed. New York: Marcel Dekker, 1982.
4. PA Schweitzer. Corrosion Resistance Tables, 4th ed., Vols. 1–3. New York: Marcel Dekker, 1995.

11
Atmospheric Corrosion of Ceramics

Ceramics are a combination of one or more metals with a nonmetal. In all probability the first manufactured ceramics date back to 6000 B.C. when clay products were produced by the burning of naturally occurring clays. The silicate glasses date back to a few thousand years B.C. It was many centuries later that porcelain ceramic products were produced by mixing 50–60% of the mullite-type clays ($3Al_2O_3 \cdot 2SiO_2$) with flint (ground SiO_2) and feldspar, an anhydrous alumina silicate containing K^+, Na^+, or Ca^{2+} ions. This mixture after firing resulted in crystalline phases bonded by a complex glassy phase. Brick-type products were the third group of traditional ceramics to be produced. Initially these were made from sedimentary clay minerals that came from deposits in oceans, lakes, and swamps. Since these clays were mixed with sand and shale they were of a high silica and alkali content. The bricks made today are vastly superior containing a glassy or vitrified bond in the product.

Traditional ceramics also include the concretes, which are composites consisting of rock, gravel, and sand, bonded together with some type of cement. Typical materials used are Portland cement or one of several polymers. Portland cement concrete and polymer concretes have been discussed in the previous chapter.

Other traditional ceramics include the abrasive cutting, grinding, and polishing group of materials and refractories, with both clay and nonclay types used for heat insulation. Silicon–oxygen–metal combinations (silicates) are found in the majority of the traditional ceramics.

The newest series of ceramics produced within the past 35–40 years have been developed on a more scientific basis. The raw materials are prepared by mechanical attrition, precipitation, sol gel techniques, plasma techniques, calcining, and spray techniques. The differences between these "advanced" ceramics and the traditional ceramics are shown in Table 11.1.

I. CORROSION BEHAVIOR OF CERAMICS

As with metals the corrosion of ceramics can take place by one or a combination of mechanisms. In general, a corrodent will attack a ceramic and form a corrosion product. Whether the reaction product is a gas or a solid will determine if the product remains on the surface or is fugitive. Reaction products may be gas, liquid, solid, or any combination thereof. If the reaction product formed is a solid it may form a protective layer against additional corrosion. When the reaction product is a combination of a solid and a liquid, the reaction layer may be removed.

The fundamental concepts of chemistry can assist in the understanding of the corrosion of ceramics. A ceramic having a basic character tends to be attacked by an environment with an acidic character, whereas a ceramic having an acidic character tends to be attacked by an environment basic in character. Also, ionic materials tend to be soluble in polar solvents whereas covalent materials tend to be soluble in nonpolar solvents, and the vapor pressure of covalent materials is generally greater than that of ionic materials and therefore covalents tend to vaporize or sublime more quickly.

Resistance of ceramics to corrosion is due to one of three basic behaviors: immunity, passivation, or kinetically limited corrosion. When a ceramic is thermodynamically incapable of spontaneous reaction with its environment it is referred to as having immunity. When the necessary thermodynamic data are available this type of corrosion resistance may be predicted by calculation. Metals, except for the precious metals such as gold, do not exhibit immunity.

For corrosion to occur without input of energy from outside the reaction, it must be both thermodynamically and kinetically possible. It is possible to calculate whether a specific corrosion reaction can occur spontaneously providing sufficient thermodynamic data are available. If on the basis of the available thermodynamic data it is determined that a given corrosion reaction cannot take place, then it is reasonable to assume that the particular corrosion reaction is not a practical problem. If the thermodynamic data predict that a given corrosion reaction can occur, it still may not be a practical problem if the reaction is kinetically limited.

The primary problem in thermodynamically predicting corrosion lies in the ability to find adequate and complete thermodynamic data and considering all the possible corrosion reactions that could be important for complex

Table 11.1 Differences Between Traditional and Advanced Ceramics

	Traditional	Advanced
Raw materials	Natural clays, SiO_2 (sand, quartzite) feldspar, anhydrous aluminum silicate, gravel, sand, kalonite	Prepared powders Al_2O_3, Si_3N_4, Z_nO_2, SiC, MgO, B_4C_3, $BaTiO_3$
Products	Whiteware, concrete, glassee, bricks, abrasives, insulators	Insulators, lasers, dielectrics, superconductors, substrates, piezoelectronics, magnetics, electro-optics, structural

ceramic compositions and structures. The use of thermodynamic data must be carefully applied and full consideration given to the impact of kinetic limitations.

The change in free energy (ΔG) as a result of a corrosion reaction is the basis for the application of thermodynamics to corrosion phenomena. Free energy can be calculated by the equation:

$$\Delta G = -nFE \tag{11.1}$$

where

n = number of electrons in the reaction
F = the Faraday constant
E = cell potential as determined by measurement vs. a reference electrode.

For a corrosion reaction to occur spontaneously the change in free energy must be negative. As a practical matter, the change in free energy is more often calculated by

$$\Delta G = (\text{free energy of products}) - (\text{free energy of reactants})$$

rather than by Eq. (11.1). For example, assume the corrosion of α-silicon carbide by hydrochloric acid to form methylene and dichlorosilyene. The reaction is

$$SiC(\alpha) + 2HCl \rightarrow SiCl_2 + CH_2$$

The free energy of formation for each of the compounds can be found in thermochemical tables. When the free energy of the products minus the free energy of the reactants is calculated, the change is +449 kJ/mol at room temperature. Since the change is positive this reaction cannot take place spontaneously at 25°C.

If we consider the corrosion of boron carbide by hypochlorous acid to form boron oxychloride and methane, we have

$$B_4C + 4HOCl \rightarrow 4BOCl + CH_4$$

Again, calculating the free energy of formation of the products minus the free energy of formation of the reactants, it is determined that the change is −1029 kJ/mol at room temperature. Since this is a negative change in free energy it is thermodynamically possible for this reaction to occur spontaneously.

Passivation of ceramics is similar to the passivation of metals in that a protective corrosion product is formed that limits further corrosion. Ceramics, as with metals, are not inert even though they have formed a protective film.

There are many ceramics that are not inert, do not effectively passivate, yet exhibit low corrosion rates. In these instances the ceramics may develop a corrosion product that is soluble in the corrodent or is not extremely protective. The slow kinetics of the ceramic–environment system are responsible for the low corrosion rate.

II. FORMS OF CORROSION

Ceramics are subject to many of the same forms of corrosion as metals. However, there are some important differences in the corrosion mechanisms.

A. Uniform Corrosion

For uniform corrosion to take place in metal, a galvanic cell must be formed. This is not possible in a ceramic because ceramics are insulators and no conductor is present—a requirement for a galvanic cell.

Uniform corrosion in a ceramic takes place as a result of chemical dissolution. The corrosion product formed is nonprotective, being either poorly adherent, soluble, or a good transport medium for the corrodent to the ceramic surface. This is a predictable form of corrosion based on test data or experience.

B. Crevice Corrosion

The mechanism of crevice corrosion for ceramics is similar to that of metals. Crevice corrosion develops in an occluded area of a corroding material. Once corrosion has started within the crevice, corrosion rates within the occluded area will increase with time as a result of either (1) depletion of a passivating component within the crevice or (2) increasing acidity within the crevice, or a combination thereof.

However, there is a difference in the mechanism for ceramics vs. metals. Increasing acidity within the crevice area of a ceramic occurs only if the ceramic is electrically conductive.

The corrosion mechanism entails the simple depletion of a passivating specie in the corrosive. This mechanism applies equally to nonconductive and electrically conductive ceramics. The passivating specie within the crevice is consumed by the corrosion reaction, making the crevice environment more corrosive than the bulk solution. Replacement of the passivating specie by diffusion is prevented by the occluded nature of the crevice; therefore, the environment within the crevice remains more corrosive. The crevice grows as corrosion proceeds forming additional corrosion product, which further restricts diffusion.

Crevice corrosion stops at this point for nonconductive ceramics. For conductive ceramics an additional mechanism occurs.

This additional mechanism causes the crevice to become more acidic and corrosive with time, resulting in increasing rates of corrosion. The mechanism may operate individually or in conjunction with the mechanism previously described. This mechanism is illustrated by the following example.

Initially, corrosion proceeds within and outside the crevice according to Reaction (11.2).

$$\mathrm{Cer} \rightarrow \mathrm{Cer}^{+} + e^{-} \quad (11.2)$$

where Cer represents any electrically conductive ceramic and Cer^{+} represents any type of cationic ceramic corrosion product. Cer^{+} reacts within the crevice to form a corrosion product [reaction (11.3)].

$$\mathrm{Cer}^{+} + \mathrm{H}^{+}\mathrm{OH}^{-} \rightarrow \mathrm{Cer}^{+}\mathrm{OH}^{-} + \mathrm{H}^{+} \quad (11.3)$$

It is a physical law that every electron produced by reaction (11.2) be consumed by a reduction reaction. Reduction of dissolved oxygen within the crevice to hydroxyl ions [reaction (11.4)] and reduction of H^{+} to hydrogen [reaction (11.5)] consumes some of the electrons.

$$\tfrac{1}{2}\mathrm{O}_2 + \mathrm{HOH} + 2e^{-} \rightarrow 2\mathrm{OH}^{-} \quad (11.4)$$

$$\mathrm{H}^{+} + e^{-} \rightarrow \tfrac{1}{2}\mathrm{H}_2 \quad (11.5)$$

By moving through the ceramic to the surface outside the crevice some electrons escape from the crevice area. They are consumed by reduction reactions in the bulk solution. A mole of H^{+} (acidity) accumulates within the crevice for every mole of electrons that leaves the crevice. The hydroxyl ion produced by the oxygen reduction initially neutralizes this excess H^+. However, the dissolved oxygen within the crevice is soon depleted. This permits excess H^+ to accumulate within the crevice accelerating the corrosion rate, which accelerates the accumulation of H^+. Restricted diffusion prevents the accumulating acidity from leaving the crevice. The crevice becomes progressively more acidic and corrosive, and corrosion rates increase progressively with time. As a result, crevice corrosion rates increase more rapidly and tend to be more severe with time for electrically conductive ceramics than for nonconductive ceramics.

C. Pitting

Pitting is a form of highly localized corrosion. Pits are the result of damage to a protective film. The damage might be the result of a corrosive attack of the protective film, or the result of mechanical abrasion of the film, or from inclusions or pockets of a susceptible phase. In the latter cases, attack takes place at the damaged site, which causes a hole to develop in the surface. Pitting can be detrimental to the function of the ceramic component since it may decrease the effective strength of the ceramic.

D. Cavitation

Cavitation per se is not a form of corrosion; however, it is closely related. Cavitation is mechanical in nature. It is the mechanical removal of material resulting from the implosion of vapor bubbles in the liquid. In so doing the protective film is removed which permits corrosion to take place, primarily in the form of pitting. There is also the possibility that corrosion will promote cavitation damage by dissolving a matrix phase that holds hard particles. Once these hard particles are loosened cavitation can take place.

E. Erosion-Corrosion

Erosion-corrosion is similar to erosion but the mechanics are somewhat different. In erosion-corrosion the fluid or moving particle removes a corrosion product on the ceramic surface. When this occurs the ceramic forms additional corrosion product, which would otherwise be protective, followed by removal of this protective product by erosion. As the cycle repeats localized ceramic loss occurs.

Ordinary erosion is the direct removal of a ceramic or component by mechanical means with no resulting corrosion.

F. Galvanic Corrosion

The general principles of galvanic corrosion of metals apply to the galvanic corrosion of ceramics. Basic requirements for galvanic corrosion to proceed are as follows:

1. There must be two electrically conductive but galvanically different ceramics in electrical contact with each other.
2. There must be a corrosive electrolyte in contact with both materials.

As with metallic galvanic corrosion, the greater the galvanic difference between the two ceramics, the greater will be the galvanic effect.

G. Selective Leaching

Ceramics consisting of multiphases are subject to selective leaching. When a phase is significantly less resistant to a specific corrodent the susceptible phase will be dissolved from the matrix while the other phases are unaffected.

Specific ions, rather than entire phases, may also be leached. What ions are dissolved is specific to the corrosive and the ceramic.

Weatherability, hardness, and other aspects of performance are affected by the leaching of small phases or specific ions.

H. Intergranular Corrosion

Selective leaching and intergranular corrosion are related but intergranular corrosion is unique in that it takes place along grain boundaries. Corrosion results from impurities and additives that tend to segregate along the grain boundaries under attack.

I. Corrosion-Assisted Cracking

Stress corrosion cracking (SCC) of ceramics is generally recognized as a type of subcritical crack growth. Relative humidity is an important factor in SCC crack growth for many ceramics. Water generally increases SCC crack growth rates.

SCC usually occurs as a result of the combined action of internal or external stresses and an aggressive environment.

J. Summary

As can be seen from the foregoing, ceramics are susceptible to the same general forms of corrosion as metals, although in some cases the mechanisms may be different. Atmospheric corrosion of ceramics may take place through any or all of these forms, depending on the specific ceramic and the contents of the atmosphere.

The same factors that promote atmospheric corrosion of metals, time of wetness, atmospheric types, initial exposure conditions, sheltering, wind velocity, and the nature of corrosion products also affect the atmospheric corrosion of ceramics.

The porosity of ceramic materials is another factor that affects their corrosion resistance. Voids present in the ceramics provide a greater surface area for attack and allow retention of the corrosion product, which results in mechanical failure due to the formation of voluminous corrosion products in the pores.

III. CORROSION OF SPECIFIC CERAMICS

The primary problem facing ceramics in relation to atmospheric corrosion is that of attack by atmospheric water vapor. This attack is time-dependent and degrades the mechanical properties of the ceramic. This form of attack is referred to as weathering and is essentially the attack by water vapor, CO_2, and SO_2. Many ceramics are immune to this form of attack at ambient temperatures.

A. Silicate Glasses

Weathering of silicate glasses takes place by one of two mechanisms. In either case condensation occurs on the glass surface. In one case the water evaporates whereas in the second case it collects to the point where it flows from the surface carrying with it any reaction products that have formed. In the former type, soda-rich films are formed that may react with atmospheric gases such as CO_2 to form Na_2CO_3. The sodium carbonate formed may act as a barrier for further weathering; however, the deposit detracts from the visual aesthetics of the glass.

If the droplets of water remain on the glass surface, ion exchange can take place with subsequent increase in the pH. Since the volume of droplets is small compared to the surface area, drastic increases in pH can take place causing severe etching of the glass surface. The rough surface formed can then collect additional solution causing additional deterioration.

Sulfur dioxide gas can react in a similar manner. However, flat glass manufacturers subject the glass to an SO_2 treatment prior to annealing of the glass. This permits the sodium in the surface layers to react with the SO_2 forming sodium sulfate, which is washed off prior to inspection and shipping. The first step in weathering is diminished due to the low alkali content of the surface.

B. Borosilicate Glasses

The borosilicate glasses are the most resistant to atmospheric corrosion of any of the glass compositions.

C. Lead Glasses

Under low-pH conditions lead will be leached from the glass. The rate of the reaction depends on the concentration gradient between the bulk glass and the acid solution, and the diffusion coefficient through the reacted layer.

D. Brick-Type Products

It is important to understand the basic difference between "common" or "face" brick and so-called "acid" brick. All bricks are produced from a clay that is primarily a mixture of alumina and silica containing small amounts of iron and other metallic oxides.

Red shale bricks are made from a clay higher in silica and iron, whereas fireclay bricks are made from a clay higher in alumina. Common red brick (usually from shale stock) and common brick of lighter color or pale brown (usually from a clay higher in alumina) after forming are fired or baked at a high enough temperature and for a sufficient time to drive out all of the water from the mud and to cement all of the clay particles together in a hard mass. All common bricks have a high water absorption (anywhere from 8% to 15%) and are leached by strong acid or alkali.

Acid bricks are produced from selected clays containing little acid-soluble components. They are fired at higher temperatures and for longer periods than the common brick. The result is a much stronger brick with a much lower water absorption, i.e., under 1% when produced from best-quality red shale and under 5% when produced from high-quality fireclay. The firing is done to create partial vitrification of the brick uniformly throughout the body of the brick, with the consequent elimination of any organic materials present. In general, acid bricks are used for floors.

Table 11.2 Atmospheric Corrosion Resistance of Selected Glasses

	Atmospheric pollutant								
Type of glass	Weathering	Chloride	SO_2	NO_x	H_2S	CO_2	UV	Ozone	Water
Sodium silicate	2	1	1			2	1		2
Alumina silicate	1	3	3	3	3	3	1		1
Borosilicate	1	1	1	1	1	1	1	1	1
Lead glasses	2	2	2	2			1		2
Fluoride glasses	1	1	1				1		1
96% silica	1	1	1				1		

1, Negligible effect; 2, some attack; 3, not recommended.

E. Red Shale Brick

In general, red shale bricks provide a lower absorption than fireclay bricks but they are more brittle. These bricks are best applied in areas where lowest absorption masonry is desired but where thermal shock is not a major factor.

F. Fireclay Brick

Fireclay bricks usually have a higher absorption than red shale bricks but they have better resistance to thermal shock. For this reason, fireclay bricks are recommended for outdoor exposures where freeze–thaw effects and extreme thermal changes can be anticipated.

Some manufacturers offer a harder burned fireclay brick with a lower absorption.

G. Other Brick-Type Products

There exist a wide variety of additional types of brick products but these are not normally used in applications dealing with atmospheric corrosion. They are used to line process vessels or used as flooring in areas exposed to severe corrosion resulting from the use of aggressive chemicals. Additional information can be found in Ref. 4.

Table 11.3 Atmospheric Corrosion Resistance of Selected Ceramics

	Resistance to:			
Material	CO_2	CO	NH_3	SO_2
Zircon	A	A	A	A
Bonded 99% alumina	A	A	A	A
Fused cast alumina	A	A	A	A
Zirconia, stabilized	A	A	E	A
Silicon carbide	B	B	A	A
Magnesite	A	B–C	A	C–D

A = no reaction.
B = slight reaction, material suitable.
C = reaction, material suitable under certain conditions.
D = reaction, material not suitable unless tested under operating conditions.
E = rapid attack, material unsuitable.

IV. SUMMARY

Ceramics are subject to the same general forms of corrosion as metals and the same factors of atmospheric corrosion that affect metals also affect ceramics. Table 11.2 lists the atmospheric corrosion resistance of selected glasses and Table 11.3 does likewise for other selected ceramics.

REFERENCES

1. RA McCauley. Corrosion of Ceramics. New York: Marcel Dekker, 1995.
2. PA Schweitzer. Corrosion Resistance Tables, Vols. 1–3, 4th ed. New York: Marcel Dekker, 1995.
3. EL Liening, JM Macki. Aqueous corrosion of advanced ceramics. In: PA Schweitzer, ed. Corrosion Engineering Handbook. New York: Marcel Dekker, 1996, pp 419–458.
4. WL Sheppard, Jr. Chemical Resistant Masonry, 2nd ed. New York: Marcel Dekker, 1982.

12 Atmospheric Corrosion of Wood

Wood is a naturally formed organic material composed of cells arranged in a parallel manner. The chemical composition of the woody cell walls is approximately 40–50% cellulose, 15–30% lignin, less than 1% mineral, 25–35% hemicellulose, and the remainder extractable matter of various kinds. Softwoods and hardwoods both contain approximately the same cellulose content.

Timber is classified as hardwood and softwood. Hardwood comes from the broadleaved trees such as oak, maple, and ash. Softwood is the product of coniferous trees such as pine, birch, spruce, and hemlock. The terms *hardwood* and *softwood* have no relation to the actual hardness of the wood. Sapwood is the living wood on the outside of the stem. Heartwood is the inner core of physiologically inactive wood in the tree. Heartwood is usually darker than sapwood.

I. NATURAL WEATHERING

Wood when exposed outdoors undergoes a number of physical and chemical phenomena. These are caused by moisture influences, sunlight, and temperature. Wood is also subject to attack by insects and fungi. All of these stress factors have an effect on the wood as shown:

Stress factor	Weathering factor
Photo irradiation	Photochemical degradation
Moisture changes	Dimensional changes, deformation, cracking, leaching, hydrolysis
Wind, rain, hail, dust, pollution	Erosion
Microorganisms	Discoloration

All of the stress factors are influenced by the following:

1. Conditions of weather exposure
2. Time factors
3. Wood properties
4. Quality of design
5. Properties and application of finish
6. Maintenance

The natural weathering process of wood creates a simple finish. The wood surfaces gradually change in color and texture, and then remain unaltered for a long period of time providing that the wood does not decay. Normally the light-colored woods will become darker and the dark-colored woods will become lighter. During the continuing weathering process all woods become gray accompanied by photodegradation and gradual loss of wood cells at the surface. Because of this, exposed unfinished wood will slowly wear away by erosion.

The weathering process is a slow-acting surface phenomenon. Most softwoods erode at an average rate of approximately 1/4 in. per century, whereas dense hardwoods erode at a rate of only 1/8 in. per century. In cold northern climates erosion rates as low as 1/32 in. per century have been reported. Western red cedar, which is a typical very-low-density softwood, may erode at a rate as high as 1/2 in. per century.

Several factors affect the physical loss of wood substance from the wood surface during weathering. These factors include not only wood species and density but also:

Growth rate
Ring orientation
Wind
Degree of exposure
Amount of irradiation
Rain action

Swelling and shrinkage stresses, caused by the fluctuations in moisture content, accompany the loss of wood from the surface. Grain raising, surface roughening, differential swelling of earlywood and latewood bands, and the formation of small parallel cracks and checks are the result. In addition, warping might occur and larger and deeper cracks might develop.

The growth of dark-colored spores and mycelia of fungi or mildew on the surface usually accompanies the weathering process. These impart a dark gray, uneven, and unsightly appearance to the wood. In such species as redwood and western red cedar, highly colored wood extractives add to the color of the weathered wood. In areas where the wood is not exposed to the sun and where the extractives are not washed away by rainwater, the dark brown color of the extractives may persist for a long time.

II. RESISTANCE OF WOOD

Wood deteriorates from two principal causes: chemical and biological attack. The chemical resistance of the wood is dependent on the ability of the wood's cell walls to resist chemical action and the extent to which the chemical penetrates into the wood.

Wood is most resistant to chemical attack in the pH range of 2–9. It is resistant to weak acids, but concentrated mineral acids tend to hydrolyze the cellulose and hemicellulose constituents. Oxidizing agents such as ozone attack the lignin in the wood which binds the fibers together. The cellulose can also be oxidized by strong oxidizing agents such as nitric acid. Aqueous solutions of nitric acid, sulfuric acid, sodium hydroxide, and hydrochloric acid can cause the most damage because swelling and degradation are simultaneous.

Decay resistance of timber varies greatly depending on the type of tree and the type of wood. Heartwood is more resistant than softwood because of the lower permeability and presence of toxic substances. Table 12.1 shows the relative resistance to decay of the heartwood of various tree species. Table 12.2 provides a more comprehensive listing. Those not shown have little or no resistance without being treated.

Wood-destroying fungi are primitive forms of plant life. They make use of the lignin (white rots) or the cellulose (brown rots) of the wood for food. The metabolic process can be halted by control of moisture, air, or temperature. Optimum conditions for growth are 20–25% moisture content, 20% of wood volume as air, and a temperature range of 75–95°F (23–35°C). Freezing of the wood does not kill the fungus; it only causes dormancy. Fungus growth can be inhibited by impregnation of the wood.

Table 12.1 Relative Resistance to Decay (Heartwood)

Trees having:	
High natural decay resistance	Moderate decay resistance
Northern white cedar	Douglas fir
Southern cypress	Eastern white pine
Western red cedar	Southern yellow pine
Redwood	Western larch
Chestnut	White oak
Black locust	Red gum
Black walnut	
Black cherry	

Table 12.2 Resistance of Wood to Decay

Softwoods		Hardwoods	
Bald cypress	E	Ash	P
Douglas fir	F-G	Aspen	VP
Hemlock, western	P	Balsa	P
Larch, western	F	Beech, American	P
Pine, ponderosa	F	Birch, yellow	G
Redwood, virgin	E	Cherry, black	E
Spruce, Sitka	F	Chestnut, American	E
White cedar	E	Cottonwood, eastern	VP
		Elm, American	F
		Elm, rock	F
		Hickory, shayback	G
		Magnolia, southern	P
		Mahogany	G
		Maple, sugar	P
		Maple, silver	P
		Oak, red	G
		Oak, white	G
		Sycamore	P
		Sweetgum	F
		Walnut, black	E
		Poplar, yellow	P

E, excellent; G, good; F, fair; P, poor; VP, very poor.

Wood is also subject to attack by insects, the most serious of which in the United States are termites of the subterranean type. These insects remove the wood internally, leaving a few signs of their activity. The cellulose in the wood is acted on by protozoa in the insect's intestine and converted to sugars for food. Control can be achieved by following construction techniques that prevent the termites from gaining entry to the structure. Since the termites live underground their entry can be prevented by the installation of metal strips between the wood and the foundation material. The U.S. Department of Agriculture has several publications available describing construction features to prevent termite entry. Termites can also be controlled by poisoning of the soil or pressure-treating of lumber with toxic chemicals.

Another insect that can cause extensive damage is the wood-boring beetle. Small surface holes and piles of sawdust are indications of the presence of this insect. Thorough surface coatings of paint or varnish are the best control measures.

III. PLYWOOD

Wood layers or plies of veneer, or veneer and lumber, in which alternate plies are laid with the grain at right angles, are given the name *plywood.* By alternating the grain direction of each ply, in adjacent plies, the two face directions are equalized in strength, stiffness, and dimensional changes. Plywood is produced as either interior or exterior type. Exterior-type plywood is designed to retain its shape and strength when repeatedly wetted and dried under adverse conditions and be suitable for permanent outdoor exposure. It is sometimes referred to as marine plywood. Exterior plywood is usually bonded with hot-pressed phenol resins.

IV. RECONSTITUTED WOOD PRODUCTS

Reconstituted wood products are produced by the formation of small pieces of wood into large sheets. The finished product can be classified as fiberboard or particleboard depending on the nature of the basic wood component.

Fiberboards are produced from mechanical pulps. Hardboard is a relatively heavy type of fiberboard that is designed for exterior exposure.

Some reconstituted wood products may be factory-primed with paint and some may even have a factory-applied topcoat. In order to protect wood from fungi and insects, treatment with toxic chemicals provides the necessary protection. A variety of materials may be used but oil and oil-borne preservatives predominate. The most widely used is coal tar creosote, which is a byproduct distilled from the coal tar produced by high-temperature carbonization of bituminous coal. It is a heterogeneous mixture of liquid and solid hydrocarbons. In order to increase penetration and reduce the coat, coal tar solutions and creosote-petroleum solutions are used extensively. Solutions as high as 50% have been used. A disadvantage to this type of treatment is the inability to apply paint.

When cleanliness and paintability is required, pentachlorophenol in volatile petroleum carriers is used. Concentrations of not less than 5% are used. These solutions are in the same price range as coal tar solutions and provide the same degree of protection.

Water-borne solutions of inorganic salts are also used. These have the advantages over the oils of greater ease of penetration and freedom from fire hazards and odor. The disadvantage is that they cause swelling and some react with metal. The primary preservative used is chromated zinc chloride. Other typical salts used are as follows:

Acid cupric chromate
Ammonial copper arsenate
Chromated copper arsenate
Chromated zinc arsenate
Copperized chromated zinc chloride

V. METHODS OF PRESERVATIVE TREATMENT

Wood can be treated with preservatives in several ways, including:

1. Under pressure in closed vessels
2. Dipping
3. Hot and cold soak
4. Diffusion

Of the four methods, treatment under pressure in closed vessels is the predominant method employed for lumber used in engineering structures.

VI. APPLIED EXTERIOR WOOD FINISHES

There are a variety of finishes that may be applied to wood that is to be exposed to the weather and atmospheric conditions. The finish selected will

Table 12.3 Painting and Weathering Characteristics of Various Woods

Softwoods	Ease of keeping painted	Resistance to weathering
Cedar	1	A
Cypress	1	A
Redwood	1	A
Pine, ponderosa	3	B
Fir	3	B
Hemlock	3	B
Spruce	3	B
Douglas fir	4	B
Larch	4	B
Hardwoods		
Aspen	3	B-A
Basswood	3	B
Cottonwood	3	D-B
Magnolia	3	B
Yellow poplar	3	B-A
Beech	4	D-B
Birch	4	D-B
Gum	4	D-B
Lauan (plywood)	4	B
Maple	4	B
Chestnut	5-3	C-B
Walnut	5-3	C-B
Elm	5-4	D-B
Hickory	5-4	D-B
Oak, white	5-4	D-B
Oak, red	5-4	D-B

1, easiest to keep well painted; 5, most difficult. A, most resistant to weathering; D, least resistant to weathering.

depend on the appearance and degree of protection desired and on the specific species of wood to be protected. In addition, different finishes provide varying degrees of protection; therefore, the type, quality, quantity, and application method of the finish must be considered when selecting and planning the finishing of wood and wood products. Table 12.3 shows the relative painting and weathering properties of various woods. The classification of paintability results from the nature of the specific wood. The higher the rating the greater the care that must be taken in applying the finish.

A. Paints

Paint coatings on wood provide the most protection because they block the damaging ultraviolet light rays from the sun. They may be either oil-based or latex-based. Either type is available in a wide range of colors. Oil or alkyd paints are borne by an organic solvent, whereas latex-based paints are water-borne. The three primary reasons for using paints are to protect the wood surface from weathering, to conceal certain defects, and for aesthetic purposes.

Paints do not penetrate the wood surface too deeply. A surface film is formed while obscuring the wood grain. Paints perform best on smooth, edge-grained lumber of light-weight species. If the wood becomes wet the paint film blisters or peels.

A nonporous paint film provides the most protection for wood against surface erosion and the largest selection of colors of any of the wood finishes. Paint accomplishes this by retarding the penetration of moisture, and reducing the problem of discoloration by wood extractives, paint peeling, and checking and warping of the wood. However, paint is not a preservative. It does not prevent decay if conditions are favorable for fungal growth. Wood preservatives are used for this purpose.

When applying the paint the surface must be clean and dry unless an emulsion paint is to be used, which will tolerate a small amount of moisture. Resins from knots in the wood should be scraped or scrubbed off with paint thinner or alcohol. The knots should then be sealed with shellac, varnish, aluminum paint, or a proprietary knot sealer. Wood being painted for the first time should have all nail holes and cracks filled with putty, after the primer has dried. All bare wood should be primed. Paint on new wood should be 4–5 mil thick. Repainting should be frequent enough to maintain a satisfactory thickness of the finish coat.

B. Water-Repellent Preservatives

Water-repellent preservatives contain a fungicide or mildewcide (the preservative), a small amount of wax for water repellence, a resin or drying oil, and a solvent such as mineral spirits or turpentine. A water-repellent preservative may be used as a natural finish. These preservatives do not usually contain coloring pigments. The type of wood determines the color of the resulting finish. The mildewcide prevents wood from darkening.

During the first few years of application the water-repellent preservative has a short life. Additional applications are usually required each year. Once the wood has weathered to a uniform color the treatments are more durable and refinishing is required only when the surface starts to become unevenly colored by fungi.

Special color effects may be achieved by adding inorganic pigments to the water-repellent preservatives. The addition of pigment to the finish helps to stabilize the color and increase the durability of the finish. Colors that match the natural color of the wood and extractives are usually preferred.

Water-repellent preservatives may also be used on bare wood prior to priming and painting or in areas where old paint has peeled, exposing bare wood. This treatment prevents rain or dew from penetrating into the wood, particularly at joints and end-grain, reducing the swelling and shrinking of wood. This reduces the stress placed on the paint film, thus extending its service life.

C. Water Repellents

Water repellents are water-repellent preservatives with the fungicide, mildewcide, and preservative omitted. They are not effective natural finishes by themselves but are used as a stabilizing treatment prior to priming and painting.

D. Solid Color Stains

Solid color stains provide an opaque finish and are available in a wide range of colors. They contain a much higher concentration of pigment than the semitransparent penetrating stains. Solid color stains totally obscure the natural color and grains of the wood. Oil-based and latex-based solid color stains form a film similar to a paint film and as such can peel loose from the substrate. Both of these stains are similar to thinned paint and can usually be applied over old paint or stains, providing the old finish is securely bonded.

E. Semitransparent Penetrating Stains

Semitransparent penetrating stains are moderately pigmented and do not hide the wood grain. They do not form a surface film, are porous to water vapor, and penetrate the surface. Since they do not form a surface film, they do not blister or peel. Penetrating stains are alkyd or oil-based and may contain a fungicide or mildewcide as well as a water repellent. Latex-based stains are also available but they do not penetrate the wood surface as do the oil-based stains.

These stains are not effective when applied over a solid color stain or over old paint. They are not recommended for use on hardwood but provide an excellent finish on weathered wood.

F. Transparent Coatings

Conventional spar or marine varnishes produce a film-forming finish and are not generally recommended for exterior use on wood. Shellac or lacquers should never be used outdoors because they are brittle and very sensitive to water. Exposure to sunlight causes varnish coatings to become brittle and to develop severe cracking and peeling, often in less than 2 years.

13 Indoor Atmospheric Corrosion

Atmospheric corrosion poses a problem indoors as well as outdoors. As can be expected, there are obvious differences in outdoor and indoor exposure conditions that lead to a difference between outdoor and indoor corrosion behavior.

Under outdoor exposure conditions the aqueous layer is influenced by seasonal and daily changes in humidity and by precipitation (dew, fog, or snow), whereas indoors the aqueous layer is apt to be governed by relatively constant humidity conditions. In this situation, for all practical purposes, there will be an absence of wet–dry cycles, and therefore the effect of indoor humidity will not introduce a time-of-wetness factor.

In general, the concentration of most gaseous pollutants found outdoors have considerably lower levels of concentration indoors, with the exception of NH_3 and HCHO. Pollutants such as HCHO and HCOOH are important indoor corrosion stimulants. They originate from adhesives, tobacco smoke, combustion of biomass, and plastics.

Another factor contributing to a decreased indoor corrosion rate is the decreased levels of indoor atmospheric oxidants, many of which are photochemically produced.

As discussed in Chapter 4, not only the concentration of pollutants but also the air velocity determines the dry deposition velocity of corrosion stimulants. Since indoors the air velocity is decreased, significantly lower dry deposition velocities will take place.

Based on the differences between the indoor and outdoor factors affecting atmospheric corrosion rates, it follows that the corrosion rate of many metals is lower indoors than outdoors. This has been verified by examining the corrosion rates of copper, nickel, cobalt, and iron in eight indoor locations. In all cases they exhibited a lower corrosion rate indoors than outdoors.

These factors do not eliminate the possibility of indoor atmospheric corrosion of materials. Designs must take into account the possibility of indoor atmospheric corrosion.

In an uncontaminated atmosphere at constant temperature, and with the relative humidity below 100%, corrosion of metals would not be expected. However, this is never the case because there are always normal temperature fluctuations; as the temperature decreases the relative humidity increases, and because of hygroscopic impurities in the atmosphere or in the metal itself, the relative humidity must be reduced to a much lower value than 100% in order to ensure that no water condenses on the surface. For all metals there is a critical relative humidity below which corrosion is negligible. These critical relative humidities fall between 50% and 70% for steel, copper, nickel, and zinc.

In the design, areas where dust particles can accumulate should be eliminated, as well as any crevices or pockets.

I. CARBON STEEL

Carbon steel can be protected indoors by lowering the relative humidity. This can be accomplished by heating the air or by reducing the moisture content of the air. In many cases maintaining the relative humidity at or below 50% will suffice. This protective measure is effective except when corrosion is caused by acid vapors from nearby unseasoned wood, or by certain volatile constituents of adjacent plastics.

The use of low-alloy steels also helps to reduce or eliminate corrosion. Atmospheric corrosion is reduced when steel is alloyed in small concentrations with copper, potassium, nickel, and chromium.

Even in indoor atmospheres carbon steel should be protected from corrosion by means of rust preventatives, painting, galvanizing, or other protective coatings, depending on the conditions of exposure.

II. ALUMINUM

Under normal circumstances indoor air poses no corrosion problems provided the temperature is relatively constant and particularly if the air is dehumidified. Products that are stored indoors in unheated buildings or trailers can be subject to staining, filiform corrosion, and other surface corrosion.

This is the result of condensation on the metal surface during cool nights after warm, humid days. Airborne pollutants, particularly SO_2, dissolve in the condensed vapor producing a conducting electrolyte. Staining problems can occur quickly even on the more resistant materials such as pure aluminum foil bonded to insulation board, alloy 5182 ends on soft drink cans, and polished alloy 6061-T6 truck wheels. If the packaging of the products permits moisture to condense on the aluminum surfaces the problem is aggravated. In the case of the beverage cans and truck wheels, they were packaged in a plastic shrink wrap that was not air-tight. They were delivered late in the day and remained in an unheated truck trailer overnight. The wrap permitted condensation but acted to prevent circulation and retarded drying. Staining usually poses a problem on products with bright surfaces.

III. STAINLESS STEEL

In uncontaminated indoor atmospheres stainless steel remains bright. Cleanliness is essential. Even the austenitic grades can rust and pit if chloride contamination is present or if dirt or deposits are allowed to accumulate on the surface.

Stainless steel is used for its ornamental as well as its corrosion-resistant properties. Quite often it is used for handrails or enclosures.

IV. ZINC

Zinc and/or zinc coatings are relatively immune to corrosion in such indoor areas as offices, shops, schools, heated warehouses, hotels, and industrial production/assembly areas where the humidity is controlled. In areas subject to occasional condensation, such as unheated buildings, sports arenas, and garages, a low rate of corrosion may be expected.

In areas that are subject to frequent condensation, such as laundries, dairies, and breweries, a significant corrosion rate will occur. In areas with frequent condensation and corrosive atmospheres a high corrosion rate will take place.

V. NICKEL

Nickel 200 when exposed indoors will remain reasonably bright and tarnish-free. In this respect it is superior to copper and brass.

VI. MONEL ALLOY 400

When exposed to an indoor atmosphere, Monel 400 produces a very light tarnish that is easily removed by occasional wiping. There is no serious problem.

VII. INCONEL ALLOY 600

Inconel 600 remains bright indefinitely in indoor atmospheres, even in damp sulfurous atmospheres that may fog nickel 200 and Monel 400. However, because of cost Inconel 600 does not find many indoor applications.

VIII. INCONEL ALLOYS 625 AND INCOLOY ALLOY 800

These alloys remain bright and tarnish-free when exposed to indoor atmospheres. As with Inconel 600, cost prevents widespread interior use.

IX. PLASTICS

Although most polymers are resistant to corrosion in indoor applications, radiation from indoor fluorescent lighting can cause yellowing in many plastics. Applications exposed to other types of artificial radiation, such as from high-intensity discharge lamps of gamma sterilization, can also subject various polymers to degradation. Prior to use the suitability of a particular resin should be checked.

X. METALLIC COATINGS

Because of the negligible time of wetness, discoloration or staining is a more important problem than corrosion. During storage or transportation zinc-coated surfaces will develop a white stain. During the rainy season it is possible for the zinc surfaces to turn cloudy overnight. The white stains on the zinc coating can grow to delicate crystals of basic zinc carbonate when moisture in the air condenses on the surface, forming a water film that remains in the surface for an extended period of time.

Aluminum coatings are also subject to staining. A white film of bayerite (ρ-$Al_2O_3 \cdot 3H_2O$) forms on the surface.

The white storage stains on zinc and aluminum coatings can be prevented via application of a chromate passivation treatment, which forms a hydrated chromium oxide film on the surface.

Sacrificial metal coatings are very effective in preventing corrosion in areas where dust tends to collect. Nickel-iron alloy coatings, developed as an alternative to nickel coatings, find application for mild interior service.

Most indoor surfaces coated with a sacrificial metal are treated again with paint or other chemicals to produce a desired surface appearance.

XI. CONVERSION COATINGS

Conversion coatings are surface treatments forming films that enhance the corrosion resistance of a surface and also increase the adhesive bonding of paint coatings. Included in this category are chromate, phosphate, oxide, and anodized coatings. More detailed information is given in Chapter 6. Because of the reduced pollution indoors the conversion coatings provide a greater degree of corrosion resistance than they otherwise would. However, phosphate and oxide coatings are still used primarily to provide a better adhesive bonding for a paint finish.

XII. PAINTING

The Steel Structures Painting Council has categorized various industrial environmental exposures or zones. Conditions in these zones are matched to industrial grade coating systems that will provide adequate protection. Interior surfaces that we are concerned with correspond to zone O and zone 1A.

Zone O refers to dry interiors where structural steel is embedded in concrete, encased in masonry, or protected by membrane- or contact-type fireproofing. Under this type of exposure painting is unnecessary. If it is desired to apply paint for aesthetic reasons or on other interior surfaces, one coat of epoxy mastic can be applied to steel, concrete masonry units, masonry block (a block filler is recommended), sheet rock (a sealer is required), wood, polyvinyl chloride, galvanized steel, and many other surfaces.

The epoxy mastics can be applied to new clean surfaces or over existing coatings of practically any type including inorganic zinc. Surfaces should be clean and free of oil residues and chemical contamination.

It is also recommended that prior to embedding the steel it be treated with an epoxy primer. The steel surface must be abrasive-blasted in accordance with SSPC-P-6 commercial blast as a minimum.

Zone 1A refers to interior areas that are normally dry (or for temporary protection). Such a zone is for mild conditions such as dry storage or office space, as well as for areas where oil-based paints presently last for 10 years or more. Even these areas might be exposed to occasional fumes or splash and spill of corrosive materials. These interior surfaces should be coated with industrial grade acrylic or epoxy mastic.

The acrylics require two coats, one of an acrylic primer and one of an acrylic finish. These paints have excellent weathering properties, low odor, good color selection and stability, and are water-based. They can be applied to steel, copper, fabric, and most other surfaces. Application can be on new clean surfaces or over existing coatings of just about any type including organic zinc. The surface must be clean and free of any oil or other chemical contamination. For best results metallic surfaces should be abrasive-blasted; for mild conditions, hand or power tool cleaning might be sufficient.

14
Designing to Protect Against Atmospheric Corrosion

When proper considerations are made at the design stage many corrosion problems can be reduced or eliminated. It is better to reduce the possibilities of corrosion in the beginning than to take elaborate steps later for the prevention of corrosion. A design engineer is not necessarily a corrosion engineer but he or she should be made aware of the factors that aggravate atmospheric corrosion so as to be in a position to guard against them.

The potential for the degradation of materials resulting from atmospheric corrosion is always present. General factors that are responsible for atmospheric corrosion are relative humidity, average temperature and temperature range, pollutant concentration, suspended dust particles, and total time of wetness. These factors are always present and should be guarded against as much as possible during the design stage because these factors themselves cannot be controlled.

Because of increasing concern about the environment, the atmosphere is generally less polluted today than several years ago. However, industrial pollution still exists in sufficient concentrations to cause degradation of materials. Industrial pollution exists in rural atmospheres that may also have routinely high humidities. In other areas mixtures of marine and industrial pollutants are common, with each type of contaminant aggravating the other. Prevailing winds will determine the distance from the shore that marine atmospheres will be found. All of these factors must be considered during the early design stages.

The total time of wetness must also be taken into account. Geographic location plays a part in the total time of wetness since the total time of wetness is affected by rain, fog, dew, snow, and dust as well as temperature.

Location is the first factor that must be considered before the design is started. This determines the type of area, whether it be rural, urban, industrial, or marine. Theoretically each of these atmospheric types is indicative of specific degrees and types of atmospheric contaminates. However, for all practical purposes this does not hold true since there is no sharp distinction between these atmospheric types. In many cases there is no "pure" urban or rural area. Depending on prevailing winds, contamination from industrial or marine areas might find its way into urban or rural areas. Because of this it is a good policy to analyze air samples from the area where the facility is to be located. The atmospheric conditions of the area should be defined in terms of temperature, humidity, and acid contaminants. Thought should also be given to special conditions that may exist in the area such as location of stacks, cooling tower drift or spray, spills, release of water or chemicals, or any other abnormal condition.

Proposed maintenance programs, frequency of inspections, and economics of upkeep should also be considered. These items will assist in selecting the proper material of construction and/or protective measures.

I. GENERAL DESIGN CONSIDERATIONS

It is impossible to completely eliminate atmospheric corrosion, but it is possible to control the corrosion rate and thereby keep it within tolerable limits. The mechanisms of atmospheric corrosion have been discussed in previous chapters of this book for various materials of construction. This chapter will deal with other aspects of the prevention and control of atmospheric corrosion and with the general principles to be followed in the application of various materials of construction, along with measures to be taken to protect them from atmospheric corrosion.

Some general design rules to be used to limit the effect of atmospheric corrosion are as follows:

1. *Avoid crevices*. Crevices provide areas where atmospheric pollutants might collect and not be washed away by rain. The collected material will continue to react with the surface and corrode it.
2. *Avoid mechanical stresses*. Excess mechanical stresses and stress concentrations increase the potential for stress corrosion cracking.
3. *Avoid galvanic contact*. Same or similar metals should be used throughout the structure. When dissimilar metals are in contact, galvanic corrosion takes place. If the use of dissimilar metals cannot be avoided the corroding part should be designed in such a way that it may be easily replaced. Fastening materials should be more noble.

4. *Avoid sheltering from rain.* Rain can be helpful in preventing or reducing atmospheric corrosion by washing away pollutants that have accumulated during a dry period.
5. *Provide corrosion allowance.* The full life expectancy of a structure may be ensured by increasing the mechanical design thickness of the member by an amount equivalent to what will be lost as a result of atmospheric corrosion during the lifetime of the structure.

A. Structures

Included under structures are many ancillary items such as ladders, grating, platforms, guard rails, and support structures. The general practice has been to use steel, galvanized iron, and aluminum for these applications. Corrosion of these items has always been a problem and has resulted in high maintenance costs.

Much time and effort has been spent over the years to provide materials that will reduce or eliminate the problems of corrosion for these applications. These efforts have resulted in the development of glass-reinforced polyester and vinyl ester materials. At the present time there are available over 100 standard structural shapes from which platforms, supports, structures, and decking can be fabricated. These materials offer several advantages including corrosion resistance, nonconductivity, high strength, light weight, and dimensional stability.

Three basic resins are used—isophthalic polyester, which is the least expensive and is used in most general applications for which it is not necessary for the material to have fire-retardant properties; isophthalic polyester resin with a fire-retardant additive that has the same general properties and area of application as the basic isophthalic resin; and vinyl ester resin, which is the most expensive but has the greatest corrosion resistance, superior strength and physical properties, and is also fire-retardant. These materials are resistant to most corrodents found in the atmosphere and to a wide range of materials that may spill on the surface. Complete corrosion resistance properties can be found in *Corrosion Resistance Tables*, 4th ed., published by Marcel Dekker.

The glass-reinforced polyester and vinyl ester resin structures are nonconductive. As such they are neither thermally nor electrically conductive. Consequently, they are ideally suited to be used as cable trays, utility truck booms, ladders, support structures for electrical devices and systems, and power junction boxes.

On a pound-for-pound basis, these fiber-reinforced plastic (FRP) materials are stronger than steel. Shapes have been used to form the superstructures of multistoried buildings and floating bridges. Joining is accomplished

Table 14.1 Typical Mechanical Properties of Fiberglass-Reinforced Standard Shapes

Mechanical property	Resin system	
	Polyester	Vinyl ester
Longitudinal direction		
Ultimate tensile strength, psi	35,000	37,500
Ultimate compressive strength, psi	40,000	45,000
Ultimate flexural strength, psi	37,500	37,500
Tensile modulus, psi $\times 10^6$	3.0	3.0
Compressive modulus, psi $\times 10^6$	3.0	3.0
Flexural modulus, psi $\times 10^6$	2.0	2.0
Ultimate sheer strength, psi	6,000	7,000
Ultimate bearing stress, psi	35,000	37,000
Izod impact (ASTM D-256), ft-lb/in. of notch sample thickness 1/8 in. except 1/4 in. for rod	30	30
Traverse direction		
Ultimate tensile strength, psi	8,000	10,000
Ultimate compressive strength, psi	17,500	20,000
Ultimate flexural strength, psi	11,000	14,000
Tensile modulus, psi $\times 10^6$	1.0	1.0
Compressive modulus, psi $\times 10^6$	1.2	1.2
Flexural modulus, psi $\times 10^6$	1.0	1.0
Ultimate sheer strength, psi	5,500	5,500
Ultimate bearing stress, psi	30,000	30,000
Izod impact (ASTM D-256), ft-lb/in. of notch	5	5
Barcol hardness	50	50
Full section in bending		
Modulus of elasticity, psi $\times 10^6$	2.8	2.8
Tensile strength, psi	25,000	25,000
Compressive strength, psi	25,000	25,000

Electrical		
Electric strength, short term in oil, 1/8 in. (ASTM D-149), V pm	200	225
Electric strength, short term in oil, kV/in.[b]	50	50
Dielectric constant, 60 Hz (ASTM D-150)[a]	5.2	5.2
Dissipation factor 60 Hz (ASTM D-150)[a]	0.03	0.03
Arc resistance (ASTM D-495), s[c]	120	120
Thermal		
Thermal coefficient of expansion (ASTM D-696), in./in. °F[c]	5×10^{-6}	5×10^{-6}
Thermal conductivity, Btu/ft^2 hr °F in.$^{-1}$	4	4
Specific heat, Btu/lb °F in^{-1}	0.28	0.28
Other		
Density (ASTM D-702), lb/in.[c]		
Solid shapes	0.065 (S-525)	0.065 (S-525)
Hollow shapes	0.065 (S-525)	0.065 (S-525)
Specific gravity (ASTM D-792)		
Solid shapes	1.80 (S-525)	1.80 (S-525)
Hollow shapes	1.83 (S-525)	1.83 (S-525)
Water absorption, 24 h immersion, ASTM D-570, max. % by wt.	0.50	0.50
Flame-retardant properties, isophthalic polyester with flame-retardant chemical additive flame resistance (FTMS 406-2023), ign/burn, s	75/75	75/75
Intermittent flame test (HLT-15) rating	100	100
Flammability test (ASTM D-635)	<5	<5
Average time of burning, s		
Average extent of burning, min	15	15
Surface burning characteristics (ASTM E-84), max.	25	25
U.L. approved (yellow card) U.L. 94 flammability	V-0	V-0
Classification		
temperature index, °C	130	130

[a]Specimen tested perpendicular to laminate face.
[b]One-inch-long specimen tested parallel to laminate face using 2-in.-diameter electrodes.
[c]Reported value measured in longitudinal direction.

by the use of fiberglass studs and nuts. The threads are standardized to be compatible with metal studs. Table 14.1 shows the typical mechanical and physical properties of fiberglass-reinforced standard shapes. Table 14.2 shows the typical properties of FRP studs.

FRP structural shapes are very light in weight, which permits them to be easily transported, handled, and set into place. They weigh 80% less than a corresponding steel shape and 30% less than a corresponding aluminum shape.

When properly designed these materials will not sag or creep under prolonged deformation or loading. Their coefficient of thermal expansion is slightly less that that of steel but considerably less than that of aluminum or wood.

Considering the foregoing properties these materials are ideal for structural supports and decking to resist atmospheric corrosion. These shapes, handrails, deckings, and such are available in color, with the color completely penetrating the material. This eliminates the need of periodic painting for protection or aesthetic reasons. This makes the maintenance needs of these materials greatly reduced over that of their metallic counterparts. Initial purchase price may be slightly higher than the purchase price for steel, but the final installed cost, including maintenance costs, makes the total installation highly competitive with less expensive materials.

Typical uses to which FRP materials have been put include:

Ladders
Support structures
Operating platforms
Pipe supports
Cable trays
Enclosures
Handrails
Walkways
Roof supports and decking
Stairs
Grating

B. Electrical Components

Electrical distribution systems in most plants consist of galvanized iron or aluminum conduit and enclosures. These materials are often subject to corrosive attack from atmospheric pollutants. Alternative materials may be used to overcome this attack. It is possible to use such items as:

Polyvinyl chloride (PVC)–coated conduit and enclosures
FRP enclosures
Stainless steel enclosures

Seals used in hazardous operating areas to prevent hazardous gases from entering the electrical enclosure (motor starter, disconnect switch, motor,

Table 14.2 Typical Mechanical Properties of Fiberglass Studs

Property	Size (UNC) 1/8-16	1/2-13	5/8-11	3/4-10	1-8
Thread shear strength using fiberglass nut in tension, lb	1250	2200	3100	4500	6500
Traverse shear on double-threaded rod, double shear, ASTM B-565, load/lb	3000	5000	7500	12,000	22,000
Traverse shear on threaded rod, single shear, load/lb	1600	2600	3800	6200	15,000
Compressive strength, longitudinal, ASTM D-695, psi	54,000	54,000	54,000	54,000	65,000
Flexural strength, ASTM D-790, psi	55,000	55,000	55,000	55,000	60,000
Torque strength using fiberglass nut lubricated with SAE 16W30 motor oil, ft/lb	8	18	35	50	110
Dielectric strength, ASTM D-149, kV/in	35	35	35	35	35
Water absorption, 24-h immersion threaded, D-570, %	1	1	1	1	1
Coefficient of thermal expansion, longitudinal, in./in. °F $\times 10^{-6}$	5	5	5	5	5
Max. recommended operating temperature based on 50% retention of ultimate thread shear strength, °F/°C		All 210/100			
Stud weight, lb/ft	0.07	0.12	0.18	0.28	0.50

Test results are for bolts with single nuts only. Properties do not apply when fiberglass is used with a metal nut.

etc.) can be used in nonhazardous areas to prevent corrosive fumes from entering the enclosures. This will reduce corrosion of the working parts.

Regardless of which material is used, enclosures and conduits should be located so that

1. Free drainage can occur.
2. Places where water and dirt can accumulate are eliminated.

This will help to prevent corrosion by reducing the time of wetness.

C. Control Rooms

Most newer plants and facilities contain control rooms, electric motor control centers, instrument control rooms, and computer centers. These are usually maintained under a slight positive pressure with makeup air being introduced from the outside. In many instances it is possible for this air to be contaminated with atmospheric pollutants such as sulfur oxides, hydrogen chloride, chlorides, hydrogen sulfide, dust, and others. These gases and dust can cause serious damage to sensitive circuitry found in computers, microprocessors, microswitches, and other delicate electronic equipment. If corrosive gases are removed from these makeup streams, costly maintenance and downtime will be prevented.

The removal of these gases and particulate matter from the makeup air stream is therefore strongly advised. It can be accomplished by treating incoming air streams in the following manner:

1. Install a filter, with a rating of approximately 25–30% ASHRE on the fan inlet.
2. On the discharge side of the fan, pass the air through an activated carbon bed, ensuring that the carbon bed is designed to remove the particular gases or vapors present.
3. After the carbon bed, pass the treated air through a final filter to remove micrometer-size particulates.

Many older control rooms and motor control centers cannot be maintained under a positive pressure because of air leakage and the inability to limit access. Under these conditions an installation similar to the aforementioned can be incorporated on a recirculating and makeup air system.

When these systems are designed, care must be taken in the selection of the construction materials for the various components and to make sure that they are compatible with the corrodents to be encountered. Installations of this type greatly reduce maintenance costs and downtime on control equipment and electrical components.

II. PROTECTION OF CARBON STEEL

Carbon steel is the most widely used material of construction. It is the primary material of choice for bridges, towers, various types of structures, and other types of outdoor constructions. In nature, iron is found in the form of various oxides, which are used as ores for refining to make steel. Steels in atmospheric service therefore tend to return to their oxide form by the process known as rusting. The rusting of steel is the most common form of corrosion that the general public sees, but it is very complex, having over a dozen variables determining the rate of corrosion. Water is the most common corrosive solvent.

Atmospheric corrosion of steel is a function of location. In country air the products of corrosion are either oxides or carbonates; in industrial atmospheres sulfuric acid is present producing iron sulfate; and near the ocean some salt is in the air producing iron chlorides. Corrosion takes place more rapidly in industrial areas because of the presence of the acid, and it is higher both near cities and near the ocean because of the higher electrical conductivity of the rain and the tendency to form soluble chlorides or sulfates, which cause the removal of protective scale.

Except in the most inocuous atmospheres, iron and steel must be protected from corrosion by one of several means. Temporary rust preventatives, galvanizing, painting, or other protective coatings may be employed depending on the severity of the atmosphere.

Weathering steels, such as ASTM standards A-242 and A-588, which are low-alloy steels of relatively high strength, form self-protecting rust films when exposed to many industrial as well as uncontaminated atmospheres. These steels contain small amounts of copper, nickel, phosphorus, silicon, manganese, or various combinations thereof which impart improved corrosion resistance. The corrosion resistance of these steels is dependent on the climatic conditions, the pollution levels, the degree of sheltering from the atmosphere, and the specific composition of the steel.

Under exposure to most atmospheres the corrosion rate becomes stabilized within 3–5 years. A dark brown to violet patina or protective film forms over this period. This patina is a rust formation that is tightly adhered to the surface and cannot be wiped off. In rural areas with little or no pollution, a longer period might be required to form this protective film. In areas that are highly polluted with SO_2 the weathering steels exhibit a much higher corrosion rate and loose rust particles are formed. Under these conditions the film provides littler or no protection.

These steels do not form this protective film in marine environments where chlorides are present. Corrosion rates will be as high as for conventional carbon steels. Table 14.3 compares the corrosion rate of carbon steel and weathering steels in various atmospheres.

The formation of the patina is dependent on a series of wet and dry periods. Periods of flushing followed by periods of drying are necessary to develop the patina. In areas where the steel is sheltered from the rain, the dark patina does not develop. Instead, a layer of rust in lighter color forms, which has the same effect.

Because the patina formed has a pleasant aesthetic appearance, the weathering steels can be used without the application of any protective coating of antirust paint, zinc, or aluminum. This is particularly true in urban or rural areas.

When employing these weathering steels, design consideration should be given to the elimination of possible areas where water, dirt, and corrosion products can accumulate. If pockets are present the time of wetness will be increased, which will lead to the development of corrosive conditions. The design should make maximum use of exposure to the weather. Sheltering from rain should be avoided.

The designer should also be aware that during the period over which this protective film is forming, rusting proceeds at a relatively high rate, during which time rusty water is produced. This rusty water may stain masonry, pavements, and the like. Precautions should be taken to avoid detrimental staining affects, such as coloring the masonry brown so that the staining will not be obvious. The weathering steels are used primarily for buildings, bridges, structures, and guard rails.

Plain carbon steel must have some form of protection applied to guard the steel from corrosion. There are many options from which to choose depending on the application and the severity of the atmosphere. As with the weathering steels, sheltering should be avoided regardless of the protective coating applied. Means of protecting include painting and a variety of metallic coatings.

A. Painting

The most widely used means of protection for outside structures including bridges, towers, and framing is that of painting. Protection of carbon steel by means of painting involves three basic steps:

1. Coating selection
2. Surface preparation
3. Coating application

The coating selection depends on the environment and what pollutants are present. This can best be ascertained by sampling the air and analyzing

Table 14.3 Corrosion of Carbon Steels in Various Types of Atmospheres

Atmosphere	Exposure time (yrs)	Average reduction in thickness (mil)		
		Carbon steel	A242 (K11510)[a] Cu-P steel	A588 (Kii430)[a] Cr-V-Cu steel
Urban industrial	3.5	3.3	1.3	1.8
	7.5	4.1	1.5	2.1
Rural	3.5	2.0	1.1	1.4
	7.5	3.0	1.3	1.5
Severe marine 25M/80 ft from ocean	0.5	7.2	2.2	3.8
	2.0	36.0	3.3	12.2
	3.5	57.0	—	28.7
	5.0	Destroyed	14.4	38.8

[a]Weathering steels.

Table 14.4 Properties of Paints

Coating type	Resistance to: Sunlight	Weather	Acid	Alkali	Moisture	Salt solutions	Comments
Vinyls dissolved in aromatics, ketones, or esters			R	R	R	R	Adhesion may be poor until all solvents have vaporized from the coating.
Chlorinated rubbers dissolved in hydrocarbon solvents	N		R	R	R		Excellent adhesion to metals, concrete, and masonry. Used on structures exposed to water and marine atmospheres.
Epoxies, polyamide plus epoxy resin	N		R	R	R	R	Harder and less flexible than other epoxies. Greatest chemical resistance of the epoxies.
Polyamide plus epoxy resin (polyamide epoxy)	N		PR	PR	R	PR	Chemical resistance inferior to that of the polyamine epoxies.
Aliphatic polyamine			PR	PR		PR	Flexible film.
Esters of epoxies and fatty acids (epoxy ester)	R	R		N			On surfaces requiring the properties of a high-quality oil-based paint.
Coal tar plus epoxy resin	N	N	R		R		Used on clean blast-cleaned steel for immersion or below-grade service.
Oil-based coatings with vehicle (alkyd, epoxy, urethane)	R	R		N			Lower cost than most coatings. Used on exterior wood surfaces.

Urethane moisture-cured	R	R	Weak R	Weak R	R	Weak R	May yellow under UV light. High gloss and ease of cleaning.
Urethanes, catalyzed		R	R	R	R	R	Expensive. Used as coating on steel in highly corrosive areas.
Silicones, water-repellent in water or solvent	R	R	N	N	R		Used on masonry surfaces.
Silicones, water-based aqueous emulsions of polyvinyl acetate, acrylic, or styrene-butadiene latex	R	R	N	N			May flash rust as a primer on steel.
Polyesters; organic acids combined with polybasic alcohols. Styrene is a reaction diluent	PR	R	R	N			Must be applied with special equipment.
Coal tar	N	N	Weak R	Weak R	R		Used on submerged or buried steel.
Asphalt; solids from crude oil refining, in aliphatic solvents	R	R	Weak R	Weak R	R	Weak R	Used in above-ground weathering environments and chemical fume atmospheres.
Zinc-rich metallic zinc in an organic or inorganic vehicle							Used as primer for steel. Zinc powders in paint provide cathodic protection to steel substrate.
Acrylic resin water emulsion base	R	R	Weak R	Weak R			

R, resistant; N, not resistant; PR, poor resistance.

to determine the corrosive conditions. Once this has been completed, a coating selection can be made. The properties of the most commonly used paints to protect carbon steel are shown in Table 14.4. All of the paints listed are not applicable for use on carbon steel exposed to the atmosphere. A discussion will follow of those formulations that are most suitable for use on carbon steel.

B. Vinyls

These are the most widely used resins for industrial coatings. The volume of solids in solution is relatively low because a relatively high amount of solvent is required to dissolve a vinyl copolymer high in vinyl chloride. As a result, most vinyl coatings must be applied in numerous thin coats of approximately 1–1.5 mil per coat. To provide adequate protection five coats may be required, making this system highly labor-intensive.

Some vinyls have been formulated to permit coats of 2–2.5 mil to be applied. However, these coatings are more susceptible to environmental and moisture penetration, thereby reducing their effectiveness.

C. Acrylics

The acrylics are exceptionally resistant to oxygen and UV light deterioration and exhibit good gloss or color retention. They can be formulated as thermoplastic resins, thermosetting resins, or a water emulsion latex.

The acrylic resins are somewhat resistant to acids, bases, weak and moderately strong oxidizing agents, and many corrosive industrial gases and fumes.

D. Epoxies

Epoxy resins by themselves are not suitable for protective coatings. When pigmented and applied, they dry to a hard, brittle film with very poor chemical resistance. Durable protective coatings can be formed by copolymerizing epoxy resins with other resins, particularly those of the amine or polyamide family, or esterified with fatty acids.

Epoxy coatings, generally crosslinked with amines or polyamines, are widely used as heavy-duty moisture and chemical-resistant coatings in atmospheric fume environments. When combined with approximately 50% of refined coal tar the amine- and polyamine-cured epoxies are one of the least water-permeable coatings available. However, coal tar epoxies are not normally used in atmospheric exposures because of their sensitivity to UV light.

E. Urethanes

Urethanes are reaction products of isocyanates with materials possessing hydroxyl groups. The polyol side may consist of a number of hydroxyl-containing materials, including water, as well as epoxies, polyesters, acrylics, and drying oils. Epoxy and polyester polyols are more chemically and moisture-resistant than acrylic polyols. They are also more expensive.

The acrylic polyol when reacted suitably to form a urethane coating is perfectly satisfactory for most weathering environments. The isocyanate can be either aromatic or aliphatic. Up to the last 15 years or so aromatic isocyanates were most commonly used for urethane coatings. Since the aromatic urethanes are very similar in resistance to the epoxy family of coatings and their cost is higher, widespread use has not taken place.

Because of the increased availability of aliphatic isocyanates this situation is changing. Although the cost of the coatings formulated with aliphatic isocyanates is approximately twice that of similar epoxy coatings and aromatic urethanes, they have superior color and gloss retention and resistance to deterioration due to sunlight. While other coatings will chalk or yellow somewhat on exposure to sun, aliphatic urethanes continue to maintain a glossy "wet" look years after application.

F. Zinc-Rich Coatings

Zinc-rich coatings owe their protective properties to galvanic protection. The preceding coatings owe their final film properties, corrosion protection, and environmental resistance to the organic composition of their constituent binder or resin.

Many of the binder systems previously discussed, particularly the epoxies, are formulated as zinc-rich coatings resulting in changes in the characteristics of the formulated coating. Urethane zinc-rich coatings are more easily applied than other zinc-rich coatings and still provide galvanic protection.

Zinc-rich coatings are subdivided as organic or inorganic. The organic zinc-rich coatings have organic binders such as polamide epoxies, chlorinated rubber binders, and urethane binders.

Inorganic zinc-rich coatings make use of silicate solutions as binders. Originally for a coating to be considered zinc-rich there had to be at least 75% by weight of zinc dust in the dry film. However, the inclusion of conductive extenders such as di-iron phosphate, to enhance weldability and burn-through, has provided equivalent corrosion protection at lower zinc loadings. Addition of other additives, such as vinyl butyrate, to inorganic formulations to form a hybrid zinc-rich formulation, together with the addition of conventional inhibitive pigments such as chromates, leads, etc., may result in equivalent performance at lower zinc loading levels, particularly when topcoated.

Zinc-rich coatings provide more effective corrosion protection to steel substrates than any other type of coating as a result of the galvanic reaction. The zinc preferentially sacrifices itself in the electrochemical corrosion reaction to protect the underlying steel. This galvanic reaction, combined with the filling and sealing effects of zinc reaction products, is what protects the steel substrate.

Zinc-rich coatings cannot be used outside the pH range of 6–10.5. Because of the high pigment content, zinc-rich coatings are relatively porous and are more difficult to topcoat than conventionally pigmented coatings. Thorough surface preparation is required before application. A blast-cleaned or pickled surface is necessary. Thicknesses of no more than 5 mil per coat are recommended.

III. COATING SELECTION

Coating selection must be based on the atmospheric conditions to be encountered. A general guideline for coating selection is as follows, based on "typical" exposure conditions:

1. *Dry interiors where structural steel is embedded in concrete, encased in masonry, or protected by membrane- or contact-type fireproofing.* Under these conditions painting is not required unless for aesthetic purposes.
2. *Exterior, normally dry, includes areas where oil-based paints last for 10 years or more.* A good choice for this area would be a latex or one-coat paint.
3. *Exterior, normally dry, includes areas where oil-based paints presently last for 6 years or more.* An oil-based paint is recommended here.
4. *Frequently wet by freshwater, involves condensation, splash, spray, or frequent immersion. Oil-based paints presently last 5 years or less.* These conditions can be handled by vinyl, coal tar epoxy, epoxy, or chlorinated rubber. The amount of UV light to which the steel will be exposed will narrow down the choices.
5. *Frequently wet by saltwater, involves condensation, splash, spray, or frequent immersion. Oil-based paints presently last 3 years or less.* Choices for this application include zinc-rich vinyl, coal tar epoxy, and epoxy.

6. *Chemical exposure, acidic (pH 5 or lower).* A vinyl or polyester paint is the best choice for an acidic condition.
7. *Chemical exposure, neutral (pH 5–10).* This is an ideal application for a zinc-rich paint.

The conditions outlined above are typical of what can be found in various atmospheric zones. No consideration has been given to conditions that could exist in a chemical plant. Protection under these conditions would have to be based on the specific pollutant.

A. Summary

Below is a summary of the performance characteristics of industrial coatings, which coupled with the previous information will assist in selecting the proper coating.

Resistance to chalking in sunlight exposure	
Aliphatic polyurethanes	Very good to excellent
Water-based acrylics	
Solvent-based acrylics	
Silicone-modified alkyds	Good
Acrylic-modified alkyds	
Vinyls	
Alkyds	Fair
Chlorinated rubber	
PVC copolymers	
Epoxies	Poor
Aromatic polyurethanes	
Polyesters	
Resistance to industrial fallout of mineral acids	
Epoxies, chlorinated rubber, vinyl chloride copolymers, some alkyds	
Resistance to industrial fallout of organic acids	
Epoxies	
Resistance to industrial fallout of alkalies	
Epoxies, vinyls, chlorinated rubber	
Resistance to condensation	
Epoxy over inorganic zinc primer	

IV. SURFACE PREPARATION

The most important process affecting the life of a paint coating system is the preparation of the surface to which the coating is to be applied. Pretreatment of the surface for paint coatings can be accomplished by chemical or mechanical processes. Chemical processes include phosphating and chromating.

Surface preparation, along with the proper application techniques of the paint system, is extremely important if adequate protection is to be achieved. Table 7.5 on page 111 provides a summary of the various recommended surface preparation specifications. The actual specifications to be used will be determined by the coating system to be applied.

It is just as important that proper application techniques be followed. Each coat must be properly applied to a controlled thickness, with successive coats applied within the proper time frame. Care should also be taken that there be no contamination of the surface between coats.

Final inspection is a must. There are available a wide range of inspection instruments for determining the thickness of each coat as well as the final total paint thickness. These include various magnetic, eddy current, and destructive thickness gauges.

Visual inspection should be made within 6 months to a year after the application, looking for inadvertent misses, thin spots, or weak areas. Any repairs, if necessary, should be made at this time.

Over a period of time the paint film will break down and deteriorate as a result of the environment. Scheduled inspections are recommended to determine the extent and rate of breakdown of the coating. Localized areas of failure should be spot-repaired before large-scale deterioration of the coating takes place.

If a scheduled maintenance approach is instituted and followed, with periodic spot touch-up and an occasional full coat applied over the entire surface, extensive costs of total surface preparation (such as complete sand blasting removal of the old coating) can be avoided.

A. Chemical Treatment

Commercial phosphating processes are proprietary in nature and in response to new demands being made on the process and changing environmental conditions the formulations are constantly being altered. Phosphating by itself does not appreciably improve the corrosion resistance of carbon steel. However, depending on the type of phosphating applied, certain properties will be imparted to the carbon steel that will improve the corrosion resistance by subsequent treatments.

Iron or sodium ammonium phosphating produces a yellow-green, violet, blue, or gray material depending on the conditions of treatment. Iron phosphating improves the adhesion of paint to the surface. It is suitable for service in moderately corrosive environments.

Zinc phosphating produces a porous coating on carbon steel. If impregnated with oil it provides protection to rock drills and weapons. It also prevents rust from spreading under defects in antirust paint coatings.

Phosphate coatings are produced by immersion, spraying, or other techniques with a phosphate solution. Zinc, iron, and manganese are typical coating materials. The prime advantage of phosphate coatings is their ability to provide a good base for paint, plastics, and rubber coatings. Phosphating converts the surface to a nonalkaline condition, protects the surface against reactions with oils in paint, and protects against the spread of corrosion from defects. Alkaline residues on the surface of a base metal cause underfilm corrosion. By improving paint adhesion through increased uniformity in the surface texture and area, phosphating extends the service life of paint films.

Chromate conversion coatings consist of soluble hexavalent chromium compounds and insoluble trivalent chromium compounds forming a film. When exposed to the atmosphere hexavalent chromium slowly leaches from the film, changing the surface appearance from irridescent yellow to either a green color or to clear. Eventually the structure of the film consists of more of the insoluble trivalent chromium compounds. The leached hexavalent chromium provides passivation for any damaged areas on the surface. In a natural atmosphere the film lasts for months or even years.

Leaching of the hexavalent chromium is the result of the chromate coating absorbing moisture. Therefore, the longer the time of wetness the shorter the life of the coating. However, the base metal is protected against corrosion for the period during which the hexavalent chromium is leaching. The life of the coating is also a function of the coating thickness. The thicker the coating the longer the service life. Pollutants in the atmosphere, particularly chloride ions, also accelerate the deterioration of the film.

In a mild atmosphere, such as indoors, chromate conversion coatings provide good corrosion resistance and surface appearance. Chromate conversion coatings also provide good bases for organic films.

V. CHROMIUM AND NICKEL COATINGS

In the natural atmosphere noble metal coatings, such as chromium, nickel, and their alloys, form passive films on carbon steel surfaces. A defect-free continuous layer of chromium provides excellent protection for a carbon steel substrate. However, deposited layers of chromium and nickel commonly have some defects, such as cracks. The corrosion rate of the carbon steel exposed at the defect would be accelerated by the galvanic action of iron and chromium. In order to prevent such galvanic corrosion the multilayer system has been developed.

The multilayer coating system makes use of the discontinuity of chromium deposits. Micro-discontinuous chromium deposits are classified into microcracked and microporous chromiums and are used as topcoats for duplex or triplex nickel coatings. Microcracks in chromium deposits greater than 0.8 μm result from internal stresses that exceed the cohesive strength of the metal. They are also induced by an undercoat of nickel. A uniform distribution of cracks is obtained by the interaction of conventional chromium (0.25–0.5 μm) and high-stress nickel (0.5–4 μm). By depositing chromium on the undercoat of bright nickel containing the fine particles of a nonelectrical conductive compound, a microporous layer of chromium is formed. Since cracks and pores penetrate to the nickel surface, a large number of microanodic sites are present on the surface. Pitting penetration to the base is delayed by the formation of many microcorrosion cells on the surface that prevent the corrosion attack from being concentrated. The number of microcracks is represented by the number of cracks that intersect a 1-cm straight line (number of cracks per cm), and the density of micropores is represented by the number of pores per square centimeter. High corrosion resistance is achieved when there are more than 100 microcracks/cm in chromium overcoats and a porosity of 40,000/cm^2 in microporous chromium deposits.

Nickel coatings are divided into three types: bright, semibright, and dull-bright nickel, the difference being in the sulfur content. Bright nickel deposits contain more than 0.04%, semibright nickel deposits contain less than 0.005%, and dull-bright nickel deposits contain less than 0.001%. The higher the sulfur content the less protective is the nickel coating. However, in a triple-layer coating system, a nickel strike of approximately 1 μ containing high sulfur (0.1–0.25%) is applied between a bright nickel and semibright nickel deposit. The corrosion potential of trinickel is the most negative of nickel deposits, so the high-sulfur layer dissolves preferentially, even when pitting corrosion reaches the surface of a semibright nickel surface. Pitting corrosion does not pass through the high-sulfur nickel layer in tunneling form since the high-sulfur nickel layer reacts together with a bright nickel layer. The pitting corrosion is diverted laterally on reaching the more noble semibright nickel deposit (see Figure 14.1).

Standards have been established for the application of chromium and nickel coatings for various atmospheres. These coatings consist of a thin topcoat of chromium and combinations of microdiscontinuous chromium and duplex or triplex nickel coatings. These will provide excellent corrosion

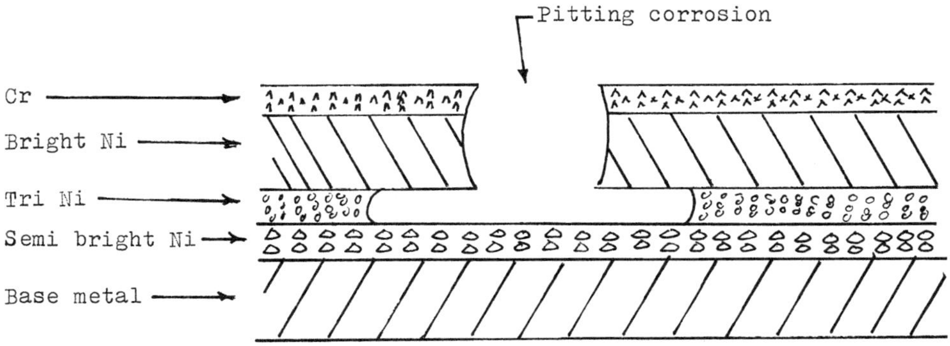

Figure 14.1 Corrosion behavior of chromium–triple-nickel coating.

resistance in natural atmospheres. The classification number is formed as follows:

Fe—designating base metal of iron or steel
Ni—designating nickel coating
Number—indicating minimum thickness (in μm) of nickel coating
Letter—designating type of nickel coating:
  b—bright nickel deposited in full-bright condition
  p—dull or semibright nickel
  d—double-layer or triple-layer nickel coating
Cr—designating chromium coating
Letter—designating type of chromium coating and its minimum thickness:
  r = conventional chromium
  f = chromium free from cracks
  mc = microcracked chromium
  mp = microporous chromium

For example, a coating on steel of 40 μm (minimum) bright nickel plus 0.3 μm (minimum) microcracking chromium has the classification number Fe/Ni 40b Cr mc.

Recommendations for coatings are as follows:

Mild atmospheres[1] Fe/Ni 10b Cr r
Moderate atmospheres[1] Fe/Ni 20b Cr r
[1]p or d nickel may be substituted for b nickel and f, mc, or mp chromium may be substituted for r chromium.

Severe atmospheres

Fe/Ni 40b Cr r	Fe/Ni 30p Cr r
Fe/Ni 30b Cr mc	Fe/Ni 25p Cr mc
Fe/Ni 30b Cr mp	Fe/Ni 25p Cr mp
Fe/Ni 30d Cr r	Fe/Ni 25d Cr mc
Fe/Ni 30d Cr f	Fe/Ni 25d Cr mp

Extremely severe atmospheres

Fe/Ni 40b Cr mc	Fe/Ni 40d Cr r
Fe/Ni 40b Cr mp	Fe/Ni 40d Cr f
	Fe/Ni 30d Cr mc
Fe/Ni 40p Cr r	Fe/Ni 30d Cr mp
Fe/Ni 30p Cr mc	
Fe/Ni 30p Cr mp	

VI. GALVANIZED COATINGS

Galvanizing or zinc coating or iron or steel will provide protection from atmospheric corrosion. There are six methods in general use to coat iron or steel as follows:

1. *Hot dipping*. Cleaned steel is immersed in molten zinc or zinc alloy. A reaction takes place forming a metallurgically bonded coating.
2. *Zinc electroplating*. Zinc salt solutions are used in the electrolytic disposition of a layer of zinc on a cleaned steel surface.
3. *Mechanical coating*. Properly prepared small steel parts are tumbled in zinc dust at room temperature. Deposition from vapor is also used to give a mechanically or physically adherent coating on steel sheet.
4. *Sheradizing*. Prepared small zinc parts are tumbled in zinc dust at just below the melting point of zinc so that a metallurgically bonded coating forms.
5. *Thermal spraying*. Grit-blasted surfaces are covered with droplets of semimolten zinc sprayed from a special gun that is fed by either powder or wire.
6. *Coatings incorporating zinc dust or flakes*. Zinc-rich paints are close in behavior to other zinc coatings.

Of the six methods available, hot-dip galvanizing is the most important. Each year more than 40 million tons of steel is galvanized in this manner using more than 2 million tons of zinc. All mild steel and cast irons can be coated in this manner but the thickness and structure of the coating will depend on

Table 14.5 Average Corrosion Rate of Zinc in Various Atmospheres

Atmosphere	Average corrosion rate (μm/yr)
Urban	2–6
Industrial	3–15
Rural	1

the alloying elements. Sheet, fabricated work, tube, and wire are coated in this manner.

The corrosion rate of the galvanized coating will vary depending on the atmosphere to which it is exposed. Table 14.5 gives the average corrosion rate of zinc in various atmospheres.

In aggressive atmospheres the corrosion rate is linear with time. When zinc forms a protective patina in milder atmospheres the corrosion rate decreases with time. The corrosion rate increases with increases in sulfur compound in the atmosphere. Corrosion also increases as the time of wetness increases. The acidity of rain affects the corrosion rate, increasing the rate if the pH is less than 5. Sheltering from direct rainfall reduces the corrosion rate.

It is better to apply the galvanizing after all fabrication and/or forming has been completed.

VII. ZINC-ALLOY COATINGS

By allowing zinc with other elements to produce a zinc alloy coating the corrosion resistance can be improved.

A. Zinc–5% Aluminum Hot-Dip Coatings

This type of coating is known as Galfan. Thickness loss of the Galfan coating is considerably less than that of galvanized steel over the same time period. In a marine atmosphere the corrosion rate tends to slow down after approximately 4 years of exposure.

B. Zinc–55% Aluminum Hot-Dip Coatings

These coatings are known as Galvalume and consist of zinc–55% aluminum–1.5% silicon. This alloy is sold under such tradenames as Zaluite,

Aluzene, Alugalva, Algafort, Aluzink, and Zincalume. Galvalume exhibits superior corrosion resistance over galvanized coatings in rural, industrial, marine, and severe marine environment. This alloy is subject to weathering discoloration and wet storage staining. To overcome these disadvantages chromate passivation procedures have been developed. Such passivation procedures also improve its atmospheric corrosion resistance.

C. Zinc–Iron Alloy Coatings

The zinc-iron alloy coatings have improved corrosion resistance over zinc in acid atmospheres but slightly reduced corrosion resistance in alkaline environments.

VIII. CADMIUM COATINGS

Cadmium coatings are commonly used on nuts and bolts. These coatings do not provide as much cathodic protection to steel as does a zinc coating. Because of this it is important to minimize defects in the cadmium coating.

Cadmium coatings are more resistant than zinc coatings to attack by salt spray and atmospheric condensation. They resist attack by strong alkalies but are corroded by dilute acids and aqueous ammonia.

Cadmium coatings should not be allowed to come into contact with food products because cadmium salts are toxic.

IX. COPPER COATINGS

Copper coatings are applied to steel for the purpose of corrosion resistance and decoration. Coating is generally accomplished by electrodeposition with a thickness of less than 15 μm.

Copper-coated steels are used as roofs, flashings, leaders, gutters, and architectural trim. The corrosion resistance to atmospheric conditions is the same as for solid copper.

X. ALUMINUM COATINGS

Aluminum coatings on steel for the purpose of protecting against atmospheric corrosion are not generally used because of cost. Their cost is much higher than that of zinc coatings.

XI. LEAD COATINGS

Lead coatings are usually applied by hot dipping or by electrodeposition. A small percentage of tin is added when the coating is to be applied by hot dipping. This is to improve the adhesion of the lead to the steel plate. If 25% or more of tin is added the resulting coating is termed *terneplate*.

Lead coatings are applied to carbon steel to resist atmospheric corrosion. Pores in the coating fill with rust thereby inhibiting further corrosive attack.

Terneplate coatings exhibit good atmospheric corrosion resistance especially under wet conditions.

XII. PROTECTION OF ALUMINUM

Aluminum finds wide applications in what would be classified as outdoor atmospheres. Corrosion of aluminum is self-limiting in all outdoor classifications whether it be seacoast, urban, industrial, or rural, and the rate of corrosion becomes very slow within 5 years except for alloys and tempers that are subject to exfoliation.

A. Seacoast Atmosphere

The most damaging to aluminum of the outdoor atmospheres is that of a seacoast atmosphere, with the severest effect being within 0.5 mile (0.8 km) of the shoreline. At a distance of 2–5 miles (3–8 km) the corrosive effect is greatly diminished. How far inland the salt mist will be carried is dependent on the prevailing wind direction and roughness of the surf action. In order to maintain a shiny metallic surface in a seacoast atmosphere it is necessary to protect the surface with a clear lacquer or an anodized coating.

The first step in the design process should be in the selection of the proper alloy for the application, both for strength and resistance to atmospheric corrosion. The 2XXX and 7XXX alloys should never be used in a seacoast atmosphere in tempers that are susceptible to intergranular corrosion, exfoliation, or stress corrosion cracking. Examples of satisfactory alloys and tempers are 2XXX–T8X and 7XXX–T7X. These artificially aged tempers have been developed to minimize susceptibility to exfoliation and stress corrosion cracking.

The higher strength alloys 5XXX and 6XXX have sufficient inherent corrosion resistance that they can be used without any protective coating. They can survive 20 years or more of exposure without any appreciable degradation, although they may darken and show a slight roughening.

B. Urban or Industrial Atmospheres

Urban or industrial atmospheres are usually regarded as those high in CO and CO_2, sulfates and sulfites, and possibly various N_xO gases, and as such are less corrosive to aluminum and aluminum alloys than are seacoast atmospheres. However, in cities using large amounts of sodium chloride (rock salt) for snow and ice removal the corrosivity to aluminum might approach that of a seacoast atmosphere during the winter months. Problems can arise when highly acidic or basic electrolytes are emitted from a nearby factory and condense on the aluminum surface.

High concentrations of N_xO gases have been found capable of causing increased susceptibility to stress corrosion cracking. In general, an urban/industrial atmosphere does not cause exfoliation; therefore, the need for protection is to

1. Retain aesthetic appearance
2. Prevent pitting perforation of thin parts
3. Avoid failure by stress corrosion cracking

One of the aesthetic problems is the gradual darkening and eventual blackening that takes place in areas where there is no regulation on smoke emissions. Other metals, masonry, and stone are affected as well as aluminum. Discoloration is common to all materials.

With appropriate protective measures practically all of the aluminum alloys provide satisfactory corrosion resistance in urban/industrial atmospheres.

C. Rural Atmospheres

Acid rain in rural atmospheres has not had as much of an effect on aluminum as it has had on steel. On aluminum alloys the acid rain causes staining and darkening of the surface. It has not resulted in more severe forms of corrosion.

Staining may be prevented by coating the aluminum alloy with a silicone-based hydrophilic solution.

D. Indoor Atmosphere

Indoor atmospheres having relatively constant air temperatures with the air being dehumidified pose no problem for aluminum or its alloys. However, storage of aluminum products in unheated buildings or tractor trailers can cause serious corrosion problems such as staining, filiform corrosion, and other surface corrosion. Condensation on the metal during cool nights following warm, humid days permits airborne pollutants, particularly SO_2, to dissolve in the condensed vapor resulting in a conducting electrolyte.

E. Protective Methods

Aluminum can be protected by either cathodic coatings or barrier coatings.

F. Anodic Coatings

Aluminum alloy products can be dipped or sprayed with an anodic coating of zinc, pure aluminum, or a more anodic aluminum alloy.

G. Alclad Coating

Alclad sheet is produced by forming a sandwich of an aluminum alloy between two sheets of pure aluminum. The aluminum sheets are placed on the top and bottom of an alloy ingot that is then rolled to form a metallurgical bond.

The barrier layer is a more resistant material than the core alloy. Its real benefit occurs when the clad surface is inadvertently scratched or otherwise damaged or if natural pitting corrosion of the cladding has reached the core. Under these circumstances the electrochemical nature of the cladding comes into play. Further corrosion spreads laterally, confining itself to the cladding rather than penetrating into the core.

H. Barrier Coatings

Although thin surface layers of more resistant metals can be applied to aluminum, these processes are not normally used because of expense and the problem of size limitations. Among the processes available are roll bonding of stainless steel and chrome plating. The most commonly used coatings are conversion coatings, anodized coatings, and organic paint films. The conversion coatings applied to aluminum are either of the chromate or phosphate types. These are used primarily for the preparation of aluminum to accept a paint coating. For environments containing chlorides, such as seacoast areas, a chromated primer should be applied first. In milder environments the paint may be applied directly to the conversion coating.

Anodized surfaces provide a more wear-resistant surface than the natural oxide, are easier to keep clean, and also provide a shinier surface. Anodized coatings are available in a variety of colors, primarily shades of gray and bronze. The color is determined by the specific alloy and the anodizing process. The thickness of the aluminum oxide coating produced in the anodizing process ranges from 5 to 30 μm, which provides ample corrosion protection from weathering and natural atmospheres. Anodized surfaces also provide an excellent base for paint.

Aluminum surfaces are painted for both protective and decorative purposes. Almost any type of paint for metals, such as acrylic, alkyl, polyester, vinyl, etc., may be used. Surface preparation is extremely important. Aluminum alloy surfaces are best prepared by the application of a conversion coating, either chromate or phosphate.

For mild environments the paint may be applied directly on the conversion coating; for more aggressive atmospheres, such as those with chlorides, a chromated primer should be applied first. The final coating is usually tailored to the application. Examples include self-cleaning paint for residential sidings and high-gloss finish for auto bodies. The topcoat is usually an epoxy or a urethane.

Clear polymer coatings are used to provide protection while retaining a glossy metal appearance. In order to provide adequate protection a hole-free finish is required. These coatings may be color-tinted to indicate that the metal has been coated or to color-code the type of coating applied. These finishes are usually applied by roller, coating large coils of sheet prior to stamping or forming of the final product.

XIII. PROTECTION OF STAINLESS STEEL

Stainless steel can best be protected from atmospheric corrosion by proper design techniques. It is essential that the design permit washing away of deposits by rain. If the stainless steel is partially sheltered deposits may be sufficiently hygroscopic and corrosive to lead to some attack that would not otherwise be encountered in the same location.

Types 304 and 304L stainless steels are normally used to resist atmospheric corrosion. The importance of design was illustrated at a location where 304 stainless steel boldly exposed surfaces were virtually unattacked, whereas in the areas partially sheltered by the roof only type 316 and 317 stainless steels remained unattacked as a result of the molybdenum content.

Cleaning on a regular basis will preserve the stainless steel surfaces. This is borne out by the "Tekko Kakan" (Iron and Steel Hall) in Tokyo, Japan. It was built in 1966, using type 304 stainless steel for the exterior curtain wall. The curtain wall is cleaned on a regular schedule to preserve the building's original beauty. Three hand-applied cleaning applications are used: water, a neutral detergent, and cleaning agents. The wall is cleaned with water every 3 months and with detergent and agent cleaners 4 times a year between the cleanings with water.

If regular cleaning is conducted type 304 stainless steel may be used in a marine atmosphere. If regular cleaning is not practical, then type 316 stainless steel must be used.

XIV. PROTECTION OF COPPER

Copper experiences a low corrosion rate when exposed to atmospheric conditions. Because of this copper has long been used for building structures such as roofs, facades, and gutters. Many copper roofs have lasted for centuries on castles and other monumental buildings.

Copper is subject to general corrosion in the atmosphere, but at a very low rate. During this exposure the surface becomes covered with copper oxides which over a period of time transform into basic copper salts, usually sulfates or, in marine atmospheres, chlorides. These basic insoluble copper salts are the basis for the green patina.

In severely SO_2-polluted atmospheres, which create acidic conditions, soluble corrosion products may be formed, thus increasing the corrosion rate. Under these conditions the green patina will not form. This condition exists near chimneys where acid smoke strikes the roof. In these areas the copper must be protected.

Rainwater running off of copper-bearing surfaces will usually contain traces of dissolved copper. In areas highly polluted with SO_2 the rainwater running off of the copper surface will contain trace amounts of dissolved copper. Such water may cause blue staining on masonry, stonework, and etc. Therefore, the rainwater from copper surfaces should be properly collected and drawn off through gutters and spouts.

Basically copper should be installed in areas where the atmosphere has either a neutral or basic character.

XV. PROTECTION OF POLYMERS

Certain polymers are subject to degradation by UV light (sunlight) unless formulated with carbon black or some other UV light stabilizer. Properties of the polymer to be used should be checked to determine whether or not it is resistant to UV light.

Many plastic formulations do not have a wide allowable temperature range. On hot sunny days it may be possible for the temperature of the plastic to be exceeded. Conversely, during the winter months temperatures might be so low that the polymer will become brittle and fracture. Local annual temperature conditions should be verified and matched to the allowable temperature range of the polymer being used.

XVI. PROTECTION OF WOOD

Wood should not be left exposed to the weather without some form of protective coating, usually paint. In addition wood that is to be in contact with the ground should be treated for protection against fungi and insects.

Various types of coatings are available to protect wood. Refer to Chapter 12 for details of the different coatings.

XVII. PROTECTION OF CONCRETE STRUCTURES

As discussed previously, one of the causes of the failure of concrete due to spalling is the rusting of the steel reinforcing (rebars). This is a particularly serious problem in bridges and roadways. It has been reported that in the United States alone 600,000 road bridges are scheduled for repair at an estimated cost of $200 billion, or 4 times the cost of their original construction. An estimated 50 million tons of carbon steel rebar is used for all manner of concrete structures each year. Early signs of corrosion of the rebar and spalling of the concrete are evident in multistory parking garages that are only 10 years old.

The primary cause is the chloride ion that results from the use of deicing salts on roadways or the presence of seawater. The chloride ion penetrates the concrete and attacks the carbon steel rebar forming rust, which expands the volume approximately 3–7 times causing the concrete to spall.

Because of the tremendous cost involved in the repair of the existing concrete bridges it is necessary to develop designs that will permit longer life of these structures. One such method is the use of stainless steel rebars that offer the following benefits:

1. *Durability (corrosion resistance)*. Stainless steel rebar has been used in several highway overpasses and parapets in Michigan, Oregon, New Jersey, Ontario (Canada), and the United Kingdom in aggressive marine environments and in the repair and renovation of historic buildings. When embedded in concrete S31600 rebar shows 5–10 times better resistance than that of carbon steel.
2. *Ease of handling and shipping*. Stainless steel rebar is easier to work with during shipment and while on site than coated rebar. The inherent protective oxide layer of the stainless steel rebar is resistant

to damage; it cannot chip, crack, or fail. It is also easily welded and can be bent into shape.

3. *Greater strength (lighter structure)*. The yield and tensile strengths of either duplex stainless steel or austenitic stainless steel rebar are superior to those of carbon steel, permitting smaller diameter stainless steel rebars to be used. This reduces the overall weight of the structure.
4. *Economical cost*. Although the initial cost using stainless steel rebar may be greater, the final cost when taking into account the cost of repairing and maintenance results in an overall saving. The estimated life of bridges constructed using stainless steel rebars is 100 years. Since stainless is so durable it is possible to use new high-strength concrete mixes such as bentonite plasticizers, superplasticizers, or polypropylene. Also to reduce costs stainless steel may need only be used in these areas of a structure where carbon steel is expected to be at high risk of corroding.

The concept of using stainless steel rebars is relatively new. At the present time no design standards or specifications are available; however, the Canadian Standards Association expects to have a stainless steel design specification available by the end of 1998 for use in Canada. Since working codes and allowable bridge loadings vary in different political subdivisions, design specifications will have to be developed by each individual political subdivision.

15
Testing for Atmospheric Corrosion

As discussed in previous chapters, there are many variables that affect atmospheric corrosion. Consequently, programs designed to study atmospheric corrosion have placed emphasis on standardization regarding test procedures. Several corrosion testing standards have been developed by the International Organization for Standardization (ISO) as follows:

1. *ISO 9223 Corrosion of Metals and Alloys.* Classification of corrosivity of atmospheres. Geneva, Switzerland 1991.
2. *ISO 9224 Corrosion of Metals and Alloys.* Guiding values for the corrosivity categories of atmospheres. Geneva, Switzerland 1991.
3. *ISO 9225 Corrosion of Metals and Alloys.* Corrosivity of atmospheres. Methods of measurement of pollution. Geneva, Switzerland, 1991.
4. *ISO 9226 Corrosion of Metals and Alloys.* Corrosivity of atmospheres. Methods of determination of corrosion rate of standard specimens for the evaluation of corrosivity. Geneva, Switzerland, 1991.

They have also implemented a worldwide atmospheric exposure program known as ISOCORRAG. Details can be obtained from:

> D Knotkova and L Vrobel, ISOCORRAG—The International Testing Program within ISO/TC 156/WG4. Proceedings 11th International Corrosion Congress, Florence, Italy, 1990, p 5.581.

This program is designed to generate a basis for the procedures used in the ISO classification system.

Two different approaches are represented in the ISO classification system to assess the corrosivity of any environment. One approach is based on SO_2 concentration, deposition of Cl^-, and time of wetness at the site, which produces an estimated corrosivity class. The other approach is based on the exposure of standard specimens of aluminum, steel, copper, and zinc for a period of a year and determining a measured corrosivity class from mass loss data. The extent of corrosion damage to aluminum, steel, copper, and zinc can be estimated once the corrosivity class has been determined by one of the above methods. The estimate can be made for short-term or long-term exposures.

Adequate data is provided by the ISO classification system for many practical purposes including prediction of long-term corrosion behavior in different environments and the need for protective coatings. Experience has shown though that certain of the observations require further classification. Among these observations include:

1. The possible role of pollutants other than those measured
2. The frequent difference in corrosion rate between the top side and bottom side of flat standard specimens
3. The frequent difference in corrosion rate between flat standard specimens and open helix standard specimens

The need for these studies is emphasized by the results of various tests conducted in which these differences appeared, e.g., the difference in corrosion rate between the top side and bottom side of flat steel plates. Table 15.1 shows the corrosion on skyward and groundward sides of steel plates at three test sites in Sweden. The atmosphere of Kvarntorp is highly polluted. Note that the corrosion on the skyward side is considerably lower than the corrosion on the groundward side. This is the result of the washing effect of the rain, which predominates in an area where the dry deposition of pollutants is considerably greater than the wet deposition of sulfur pollutants.

Table 15.1 Corrosion on Skyward and Groundward Sides of Steel Plates at Three Test Sites in Sweden

	Corrosion rate ($g/m^2/yr$)	
Test site	Skyward	Groundward
Kirima	73	59
Bohus Malinon	145	107
Kvarntorp	279	466

At the two other sites, which have less pollution, the situation is reversed and the corrosive effect of the rain predominates.

One gaseous pollutant that is of concern is NO_2. Environmental concerns and regulations have had the effect of reducing the quantity of SO_2 being emitted to the atmosphere. However, it is estimated that the emission of NO_2 in these same areas over the past decade has shown a continuous increase. This was reported by RB Huser in *Acid Deposition: Long-Term Trends* (National Academy Press, Washington, D.C., 1986, p 48).

Laboratory studies using synthetic air have provided evidence of increased corrosion rates when NO_2 is added to air containing SO_2. The synergistic effects of SO_2 and NO_2 interactions are based on observations of several metals:

Copper

P Eriksson. Effects of SO_2 and NO_2 on metal surfaces; atmospheric corrosion of copper and surface reactions on gold. Thesis, Chalmers, University of Technology and University of Gothenburg, Goteborg, Sweden, 1982.

Nickel

S Zakipour, C Leygraf, G Portnoff. Studies of corrosion kinetics on electrical contact materials by means of quartz microbalance and XPS. *J Electrochem Soc* 133:873 (1986).

Steel

J Henriksen. Corrosion rates of various metals in SO_2/NO_2 polluted atmospheres. Proceedings 10th Scandinavian Corrosion Congress, Swedish Corrosion Institute, Stockholm, Sweden, 1986, p 39.

Zinc

J E Svennsson, L G Johansson. A laboratory study of the effect of ozone, nitrogen dioxide, and sulfur dioxide on the atmospheric corrosion of zinc. *J Electrochem Soc* 140:2210 (1993).

An international exposure program within the U.N. Economic Commission for Europe (UN/ECE) was implemented in 1987 with the general aim of performing a quantitative evaluation of the effect of sulfur pollutants in combination with NO_2, other pollutants, and climatic parameters. Thirty-nine test sites in 12 European countries, the United States, and Canada were included. The program is based on exposure of structural metals, stone materials, paint coatings, and electric contact materials at test sites where measurements of environmental parameters are already in progress. Results were reported by:

V Kucera, A T Coote, J Henricksen, D Knotkova, C Leygraf, C Reinhardt. Effect of acidifying air pollutants on materials including historic and cultural monuments—an international exposure program within UN ECE. Proceedings 11th International Corrosion Congress, Florence, Italy, 1990 p 2: 433.

Analysis of the results after 1 and 2 years exposure indicates that the influence of SO_2 on the corrosion rate of carbon steel, weathering steel, zinc, bronze, sandstone, limestone, and nickel is significant. Conversely, no influence of NO_2 has so far been observed in any of the materials studied. This discrepancy between the laboratory exposures and the field tests must be investigated. Perhaps as a result of catalysts or oxidizers in field exposures, which aid in promoting the oxidation of S(IV) to S(VI), the effect of NO_2 was hidden.

Mansfeld and Vijayakumar conducted another exposure program in southern California in areas having very low levels of SO_2 (<10 ppb). Under these conditions SO_2 is shown to be of little consequence whereas NO_2, which is photochemically converted to HNO_3, is of greater importance. The results were reported in *Corrosion Science* 28: 939, 1988.

There is sufficient laboratory evidence that synergistic effects caused by interacting atmospheric constituents are significant. However, to what extent they contribute during field exposures remains to be proven. In addition to SO_2 and NO_2, the interaction between SO_2 and O_3 and between H_2S and Cl_2 is expected to be important (1,2).

I. TESTING TECHNIQUE

The most reliable corrosion testing technique is to expose actual parts in their intended environment and evaluate the resulting corrosion. Reliable corrosion data can be obtained if the corrosion tests are designed and conducted properly. The simplest techniques involve determination of a change in dimension or weight and observation of the corroded surface.

A. Dimensional Change

One of the most straightforward corrosion testing techniques is ultrasonic thickness measurement. The thickness of the test piece is measured at the beginning of the test period and after regular intervals. The difference between thickness measurements can be divided by the time interval to obtain the corrosion rate. Changes in thickness should be plotted against time so that changes in corrosion rate can more easily be traced to changes in environment. This method is not accurate enough for laboratory testing but is suitable for field testing.

B. Microscopic Examination

When testing for dealloying it is often necessary to examine a polished cross-section to determine how much metal remains. In this type of corrosion considerable damage can take place because of the removal of some of the elements from the solid alloy. The amount of damage that has been done will not be proportional to a change in external dimensions or in weight. It is necessary to prepare a cross-section of corroded metal for metallographic examination and determine microscopically how much metal remains, as shown in Figure 15.1.

C. Weight Change

The most widely used testing technique to determine corrosion rates is that of weight loss. More materials can be evaluated for a given amount of money with weight loss testing than any other technique. In weight loss corrosion testing a prepared specimen is exposed to the atmosphere for a period of time and then removed to determine how much metal has been lost. The corrosion rate can be calculated by taking into account the exposure time, weight loss, surface area exposed, and the density of the metal.

Corrosion rates may be calculated using the equation:

$$\text{mpy} = \frac{\text{WL} \times 22.273}{DAT} \tag{15.1}$$

where

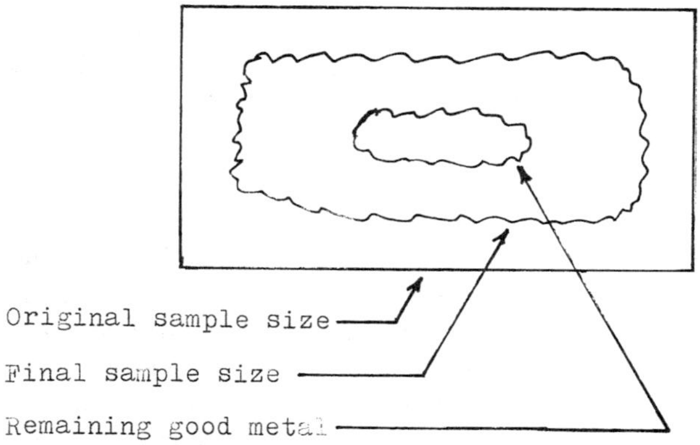

Figure 15.1 Microscopic corrosion measurement is often used to determine how much good metal remains.

mpy = mils per year
WL = weight loss, g
D = density, g/cm^3
A = area, in.^{2}
T = time, days

The weight loss should be measured to the nearest milligram (0.0001 g). Densities for most alloys can be found in either the *Metals Handbook* (3) or *Properties of Some Metals and Alloys* (4). Dimensions should be measured with sufficient accuracy that the area can be calculated to ±1%. The area of *all* exposed, including thin edges, should be included in the A term. Time should be measured to an accuracy of ±1% of the total exposure time.

D. Specimen Preparation

When corrosion testing is being conducted to determine which metals are best suited for the application, it is best to test them with a surface condition that is representative of how they will be used in practice. For this, "as-rolled" surfaces with sheared edges should be used as corrosion coupons. If the metal being tested is susceptible to stress corrosion cracking, then cold-worked or stressed specimens should be considered.

E. Specimen Holders

The test specimens must be physically separated from each other and supported on inert racks. When supported on metal racks the test specimens must be insulated from the metal rack and from each other to prevent galvanic action. Atmospheric corrosion coupons are usually held with ceramic electrical insulators as shown in Figure 15.2. The coupons are usually mounted either at an angle of 20–30° from the horizontal or vertically (see Figure 15.3). Atmospheric corrosion tests usually last at least 1 year and very often extend to 3–5 years.

F. Specimen Cleaning

Before being cleaned the specimens should be thoroughly examined using a binocular microscope when possible. The location and type of deposits and localized corrosion should be recorded and photographs taken if of value. Samples of deposits and corrosion products should be removed for chemical analysis.

The corrosion products should be removed completely from the coupons using an appropriate cleaning solution (5) and they should be weighed to determine weight loss. The corrosion rate is calculated using Eq. (15.1).

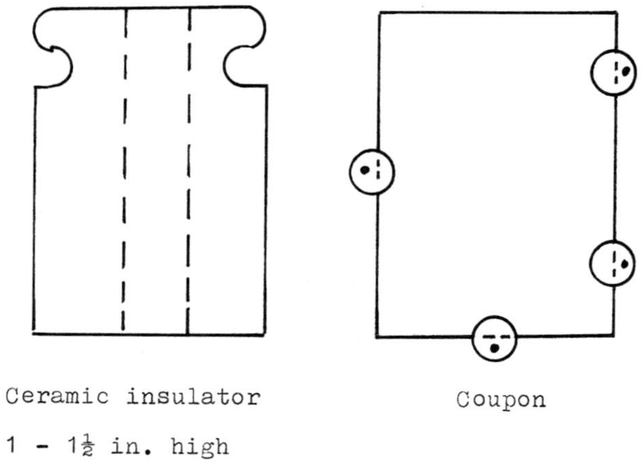

Figure 15.2 Ceramic electrical insulator supporting atmospheric corrosion coupon.

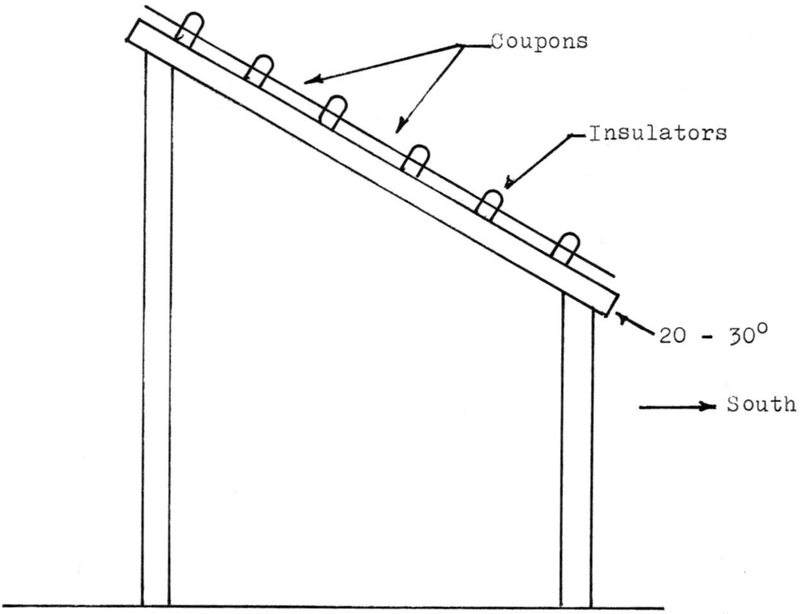

Figure 15.3 Atmospheric corrosion test rack.

II. TESTING FOR POLYMER COATING ADHESION

Regardless of the excellent properties a coating may possess in terms of weathering, such a coating is useless if it does not also possess good adhesion properties. In chemical terms there is a similarity between paints and adhesions or glues. Both materials are basically organic coatings. In either case adhesion is the property required, though with different emphasis on its intensity depending on its intended use.

Such coatings are basically polymers consisting of more or less cross-linked macromolecules and a certain amount of pigments and fillers. Substrates for the coating may be various metals, woods, plastics, paper, leather, concrete, or masonry.

It should be noted that these substrate materials might inhibit a rigidity higher than that of the coating. Under these conditions fracture will occur within the coating if the system experiences external force of sufficient intensity. Cohesive failure will be the consequence if the adhesion at the interface exceeds the cohesion of the paint layer. Otherwise, adhesive failure is obtained producing a definite separation between coating and substrate (see Figure 15.4).

In practice both types of adhesion damage are encountered. Cohesion damage indicates an optimum adhesion strength and that improvement of the inherent strength of the coating material is required.

There are three basic causes of adhesion failure:

1. *Regular or mechanical stress*, affecting the bond strength at the interfaces as well as the bulk of the material. Stress may be tensile stress (perpendicular to the surface) or shear stress (appearing along the plane of contact).
2. *Thermal changes*. If the coefficient of expansion between the substrate and paint film is different, the possibility exists that interface stress conditions of such a high intensity can be produced that the paint film becomes detached from the substrate.
3. *Chemical effects* may be produced as the result of a chemical penetrating through the coating, becoming absorbed at the interface, and causing loss of adhesion.

These three effects should be kept in mind when measuring adhesion because the method of testing should reproduce the end-use conditions.

A. Measuring Adhesion

The results obtained from an adhesion test, measure the maximum mechanical stress that can be obtained at the interface, are expressed as force per

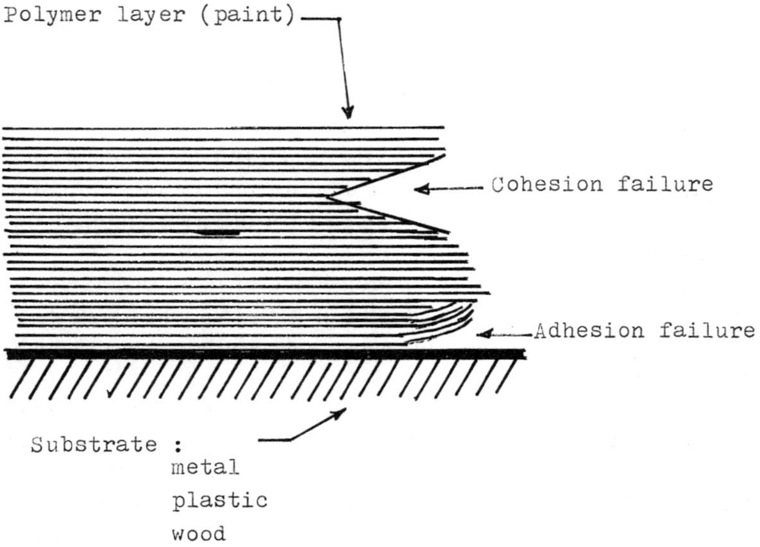

Figure 15.4 Cohesive and adhesion failure.

unit area, and are specified as either tensile or sheer stress. The energy expended in breaking the bond at the interface is expressed as work of adhesion and is equivalent to the product of adhesion strength and the distance between the separated surfaces of coating and substrate immediately after detachment. This quantity has the dimensions of force per unit length and is exactly the value obtained with the peeling test.

B. Standardization of Adhesion Tests

As discussed, the degree of adhesion is critical to the effectiveness of a paint coating. Consequently, the specification of the degree of adhesion must be supplied by each paint manufacturer. Because of this, routine methods for the measurement of the degree of adhesion have been established for the purpose of quality testing. The most popular test used is the cross-cut test, which is quite often complemented with pull-off methods. Both methods have been the subject of standardization.

C. Cross-Cut Test

The ISO has established procedures for conducting this test in ISO 2409 Paint and Varnishes Cross-cut Test. Basically, this test consists of cutting a

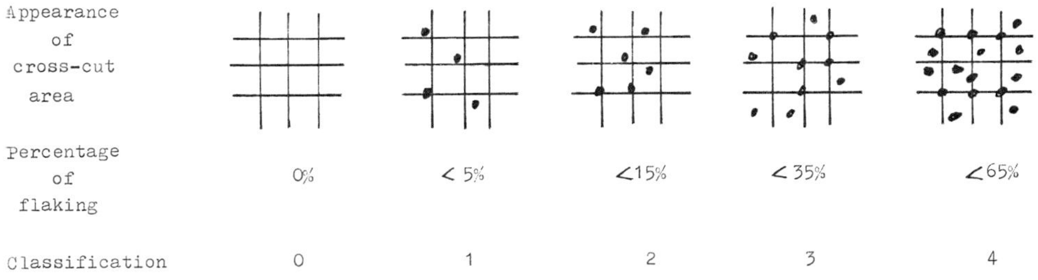

Figure 15.5 Classification of paint film in cross-cut adhesion test.

lattice pattern into the paint film penetrating through to the substrate. A multiple cutting tool, consisting of a set of six "knives" 1 or 2 mm apart, is used to produce a uniform pattern.

The test results are evaluated by estimating the amount of paint flakes separating from the substrate. This results in the paint being assigned a classification number as shown in Figure 15.5. It is permissible to brush off the loose parts or to remove them using adhesive tape.

Class 0 indicates perfect adhesion. Class 1 or class 2 should be interpreted as an unsatisfactory result. All higher classes would indicate a need to improve the adhesive quality of the paint.

D. Tensile Methods

Procedures for this test method are covered in ISO 4624 Paint and Varnish Pull-off Test for adhesion. In preparation for this test a stud, normally made of steel, is glued with the coating and is subjected to axial tension until detachment of the paint film takes place. The result is a maximum tensile stress that is possible at the interface (see Figure 15.6). The maximum shear stress is obtained when a torque is applied about the axis of the stud.

The value of adhesion strength obtained from either method is of the same order of magnitude (6). Results using the torque principle tend to give results of cohesive failure with lower results for adhesive failure.

Results obtained from these tests cannot be considered to produce material constants for the bonding components because other factors such as temperature, speed of deformation, and form or size of the stud will have an influence on the results.

E. Delamination Tests

In the previous tests the primary concern was stress conditions at the interface. Delamination is the result of peeling forces that attack the bond be-

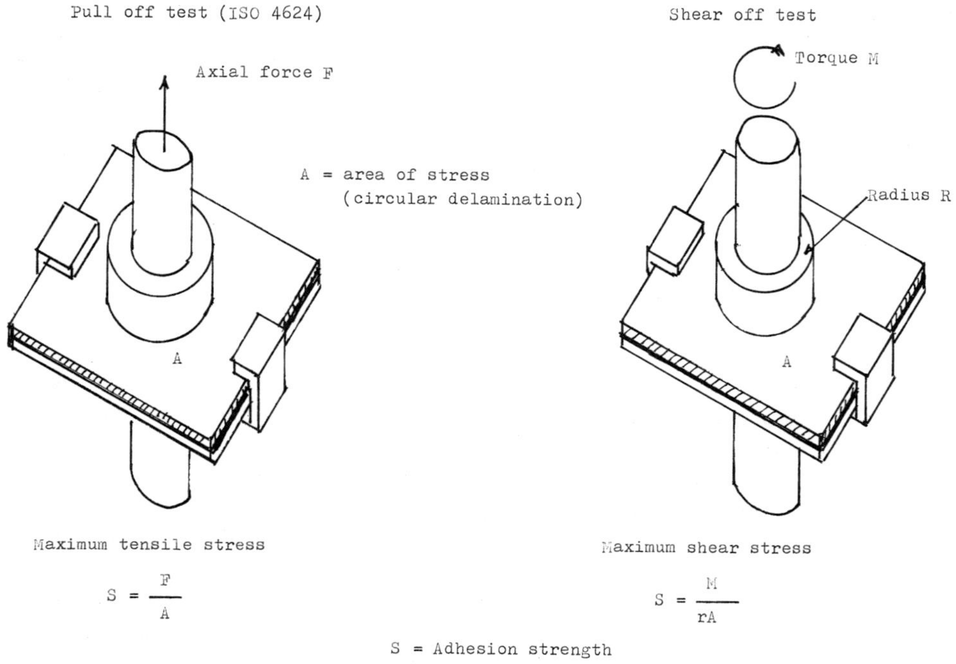

Figure 15.6 Tensile and shear method for determination of adhesive strength.

tween the paint film and substrate. The attack occurs either at well-defined lines, so that the coating is detached in strips, or at one point only, causing delamination to progress radially in the form of a blister. Both conditions occur in practice.

Three types of tests have been developed to test for delamination.

F. Knife Cutting Method

In this method film separation is obtained by means of a sharp knife pushed along the interface with a measured force (7). The process of detachment comprises both tensile and shear stresses, which eventually causes detachment of the film. In this method a particular system of forces becomes effective. It is determined by rake angle of the knife, coating thickness, friction between cutting tool and coating as well as substrate, amount of energy stored elastically in the film and energy losses caused by plastic deformation, fracture energy occurring during decomposition within the film, and other effects of minor importance.

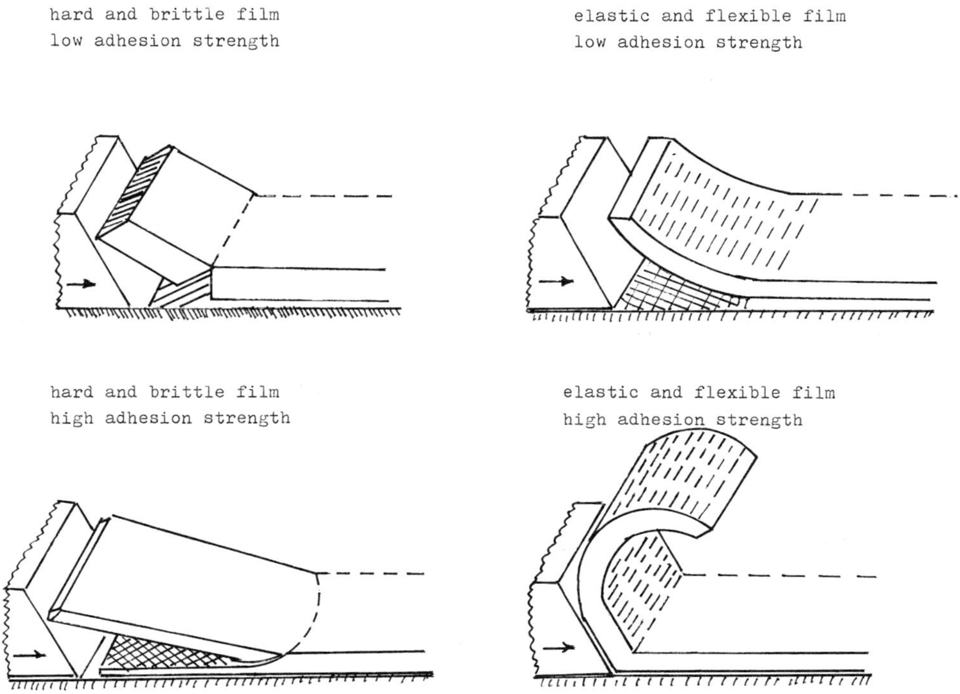

Figure 15.7 Influence of mechanical paint film properties on the results of knife cutting test.

All of these factors must either be strictly controlled or their influence estimated as accurately as possible. As shown in Figure 15.7, the details of the separation process indicate how much importance can be attributed to the various factors.

G. Peel Test

Quite often a coating fails by gradual peeling either from a scarcely covered edge or from a line-shaped damaged area. Because of this it is appropriate that a suitable test procedure be utilized to investigate this condition.

Detachment of the film is the result of the combined effects of both positive and negative tensile stresses and shear stress. To conduct the test a strip of sufficient width is marked on the sample by two parallel cuts of adequate length. As shown in Figure 15.8, the angle under which the load is applied has a bearing on the results (8). The details of the separation

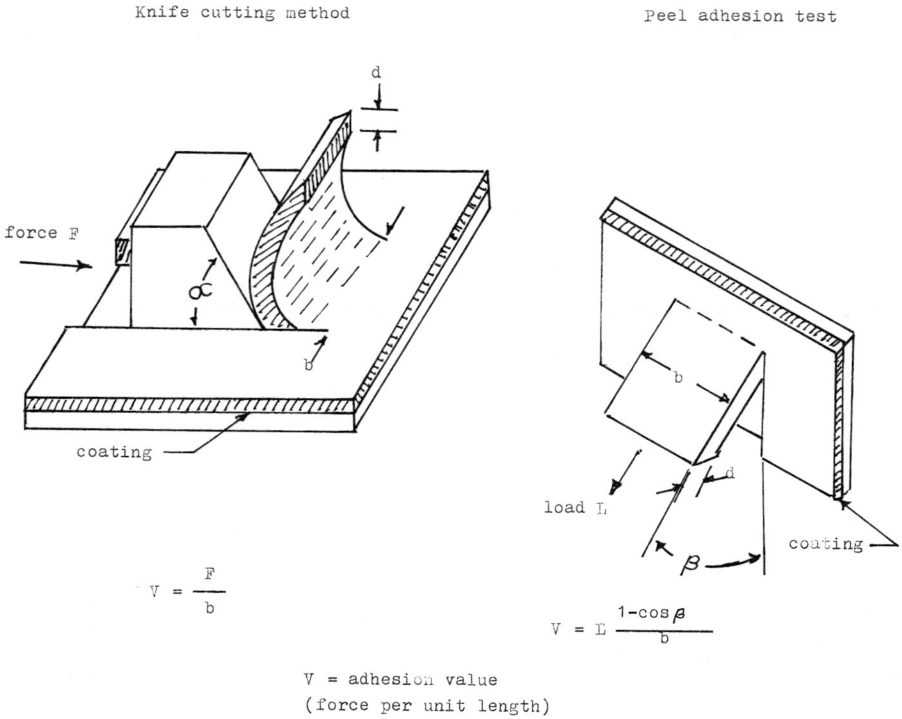

Figure 15.8 Devices for measuring adhesion on the basis of delamination procedures.

process are quite complicated although at first glance the process might appear relatively simple (9).

The viscoelastic nature of the film is a contributing factor and this property is also affected by the amount of pigmentation in the film (10). Because of the viscoelasticity of the film the test results also depend on the velocity of the detachment process.

Although the peel test is a practical method the results cannot be interpreted in terms of the bonding mechanism.

H. Blister Method

The first indication of deterioration of any paint film used for corrosion protection is the formation of a blister (10). Consequently a method was developed to investigate the behavior of such a coating material under conditions that may lead to film detachment in the form of a blister.

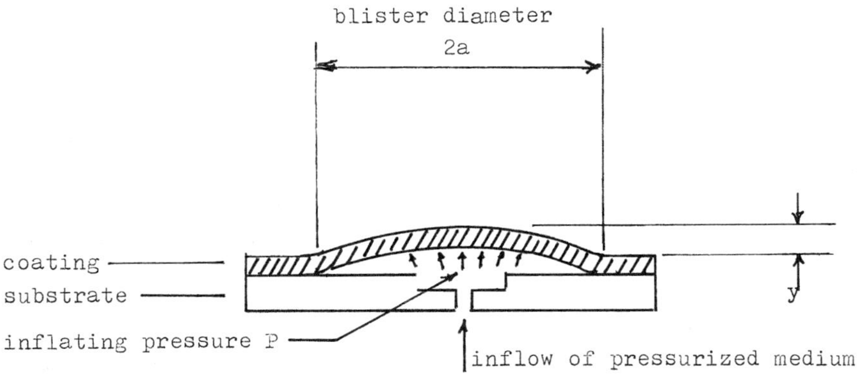

Figure 15.9 Blow-off test to measure adhesion based on blister dimensions and pressure.

The test is conducted by first boring a hole in the substrate at the site where a blister is to occur, then plugging with a material such as Teflon on which there will be no adhesion during painting. The substrate is then painted in the normal way (see Figure 15.9).

Hydrostatic pressure is supplied in the hole, either with compressed air or with a fluid (oil, mercury, etc.). The pressure is the primary measure of the progress of the debonding process. The diameter and height of the blister will determine the work of adhesion. From these data together with the tensile modulus of the film and its thickness a critical pressure value can be calculated:

$$P = 4.75E \frac{dy^3}{a^4}$$

It is this process that causes the blister to grow; therefore, this pressure serves as the basis for determining the adhesive strength (W).

$$W = 0.65Pd$$

I. Debonding from Mechanical Damage

Mechanical damage in the form of impact deformation or scratches not only affects the appearance of the film; loss of adhesion can also result. Tests have been developed to determine the resistance of a paint film from these occurrences.

J. Impact Tests

Impact tests are designed to test the adhesion of a coating providing protection of a metal substrate against corrosion when impacted by solid objects such as stone chips. The condition experienced in the field can be duplicated by means of a steel ball impinging on the test piece (Figure 15.10). As a first approximation the transfer of forces through the film is equal to the case of static loading, which means that it can be calculated in basically the same way as for an indentation test.

In the debonding area two types of stress are present: negative tensile (compressive stress) in the center of the circular detachment area and, still more important, shear stress in the annular region. The maximum diameter of the debonding area can serve as a measure of the adhesion at the interface. The larger the diameter the lower the adhesion rate (11).

Theoretically it is possible to calculate the value of adhesion in absolute terms, but there are other factors that must be taken into account. First is the influence of film thickness. The detachment area of a thin film is more concentrated in the immediate area of the impact, whereas areas with thicker

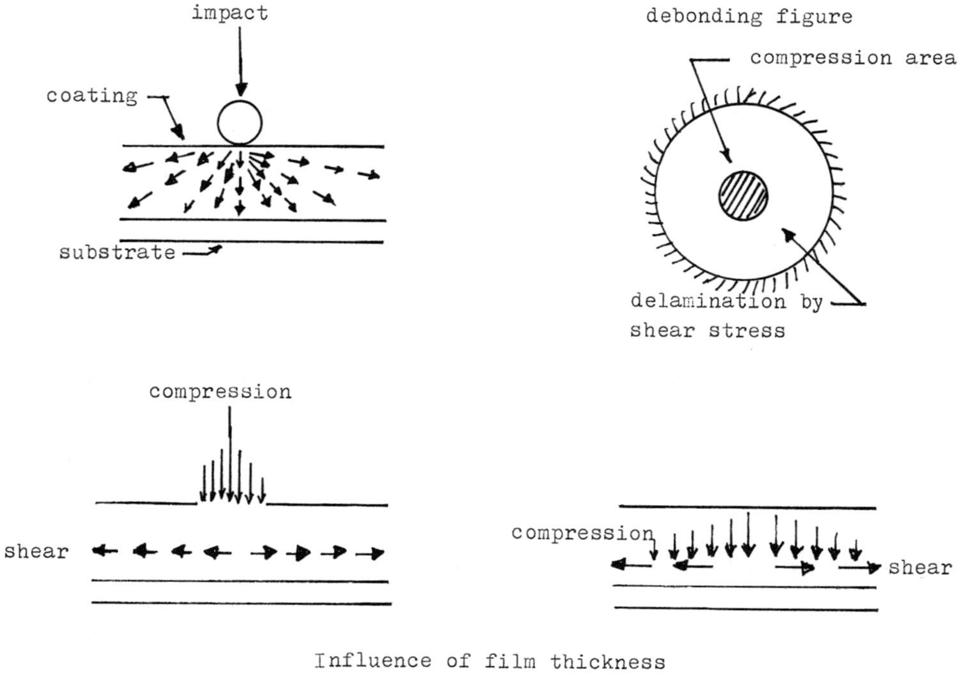

Figure 15.10 Effect of impact loading of a paint film on adhesion.

coatings, under the same test conditions, will have a much larger area affected by stresses leading to delamination. Other factors that must be taken into account relate to the mass and diameter of the steel ball, the speed before and after impact, and the time of impact (usually very short) in which the energy is being transferred to the film.

K. Debonding from Indentation

When a needle-like object is perpendicularly pressed into the surface of a coating that has been bonded to an undeformable substrate, most of the deformation will take place within the film. However, there will also be a certain debonding effect at the interface.

As shown in Figure 15.11, a peeling moment may be calculated which may serve as a measure of the film's capacity to withstand delamination in the area of the indentation site (12).

It is not necessary to use a needle-like indenter. Other shapes can be used successfully. Experience has shown that an indenter having a 60° angle cone will provide results that permit calculation of adhesion values (13). This test has the advantage of providing values for the bond strength in absolute terms as well as information about the durability of the connection between the coating and substrate under these specific loading conditions.

III. DETECTION OF FLAWS

Once a coated substrate has been placed into service it is important to periodically inspect the film for defects. It is just as important to know whether or not there has been any deterioration in the bond strength between the film and the substrate. Methods to accomplish this have been developed based on acoustic or thermo-optical principles.

A. Ultrasonic Pulse-Echo System

Monitoring of ultrasonic echoes is widely used to detect and identify irregularities in materials that are supposed to be flawless. This method may also be used to determine the quality of bonding between a polymer and a substrate.

As seen in Figure 15.12, the incident of ultrasound will be partly reflected and transmitted at each interface of the test piece, including the backing. When the film is completely adhered at the interface the amplitude of the reflected pulse will be relatively low in contrast to the reflected pulse that travels through the substrate and is reflected on its free boundary, ap-

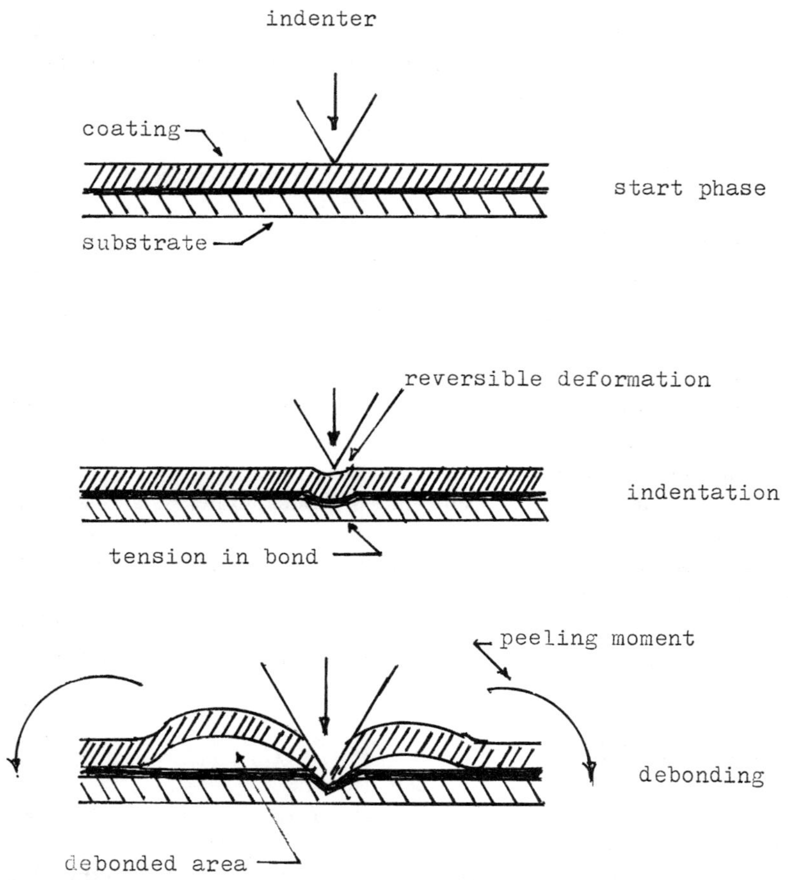

Figure 15.11 Principle of measuring adhesion at the polymer–substrate interface resulting from indentation damage.

pearing later in the sensor. When there is a defect at the interface containing air or when the joint is disbonded the amplitude of the related ultrasonic pulse increases considerably because of the very low acoustic impedance at that location (14).

B. Acoustic Emission Analysis

Acoustic emission analysis shows promising potential for the detection of flaws in coating systems. However, considerable additional information is

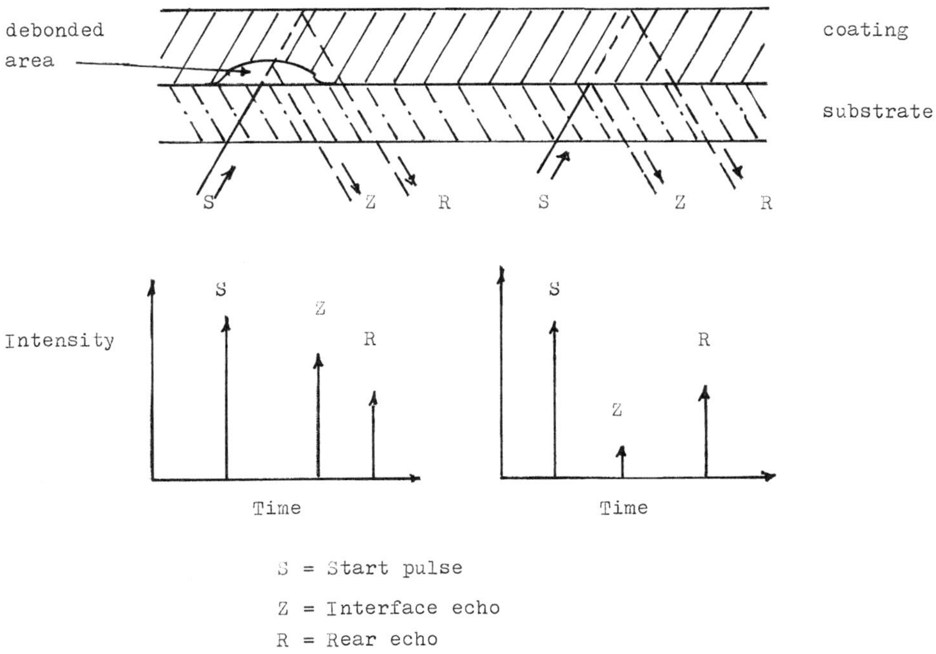

Figure 15.12 Ultrasonic pulse-echo technique for detection of adhesion flaws.

required before the primary results that have been obtained can be interpreted correctly (15,16).

C. Thermographic Detection of Flaws

Thermographic detection methods possess the advantage of yielding a remote mapping of the distribution of temperature on the surface. This principle has been applied for the detection of flaws.

As the heat flux proceeds from the substrate toward the surface of a test piece any flaw or debonded area will produce a difference in the otherwise uniform temperature on the surface. This is illustrated in Figure 15.13. An infrared sensor is used to make visible the shape of the area in which the temperature decreases at the surface, which is an almost exact reproduction of the real defect. The thinner the coating the better the agreement between the real shape of the disbonded area and the image on the surface. The process is similar to the formation of a shadow from a distant object.

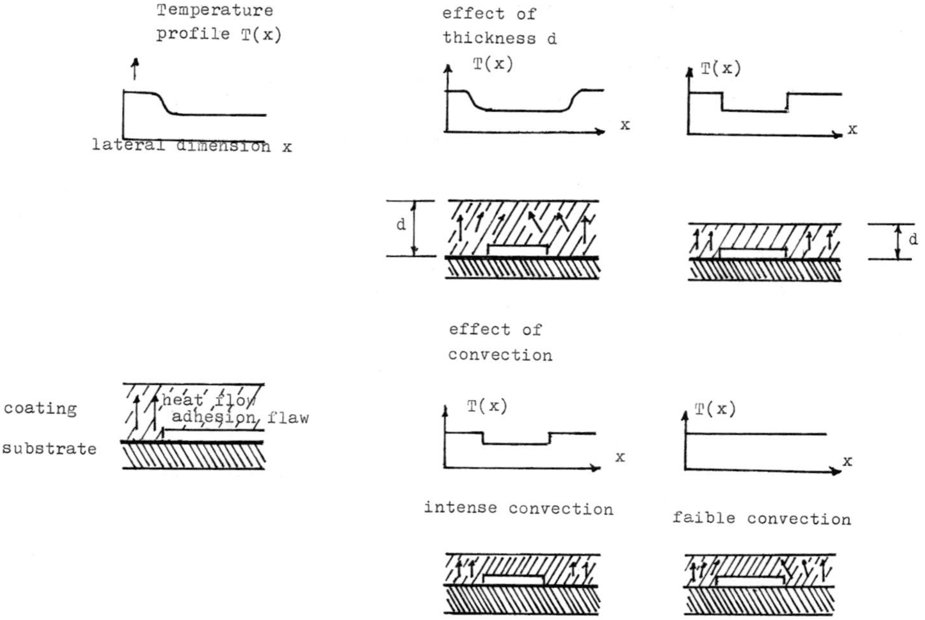

Figure 15.13 Thermographic detection of debonding.

A distinct reproduction of the defect on the surface can be expected if the linear dimension of the debonded area is twice the distance from the interface to the surface.

Additional details can be found in Ref. 17.

IV. MONITORING CORROSION UNDER PAINT FILMS

Wolstenholme (18) summarized corrosion potential measurements and their applicability to coated metals. In general, it can be assumed that movement of the corrosion potential in the noble direction is indicative of an increasing cathodic/anodic surface area ratio and is indicative that oxygen and water are penetrating the coating and reaching the metal–coating interface. Movement of the corrosion potential in the active direction indicates that the anodic/cathodic surface area ratio is increasing and the overall corrosion rate is becoming significant. The increase of positive potentials with time indicates that alkaline conditions caused by the oxygen reduction reaction are developing rapidly at the metal–coating interface and delamination is of concern. In the case of steel, an increase in active potentials indicates rusting beneath the coating and is a warning that the coating life is limited.

These measurements provide a means to monitor coatings in the field and permit steps to be taken before serious damage results due to atmospheric corrosion. Early preventative maintenance is less costly than extensive repair work resulting from a failed coating.

V. INDOOR CORROSION TESTING

As discussed previously, the ISO classification system for predicting outdoor corrosion rates was based on SO_2 concentration, deposition of Cl^-, and time of wetness. These factors have a lesser influence on indoor corrosion rates. A similar classification system has been developed by the Instrument Society of America (ISA) for the prediction of corrosivity in indoor environments (19). This standard includes measurements of SO_2, NO_2, H_2S, and Cl_2 in addition to relative humidity and temperature.

Analysis of corrosion products on copper, zinc, nickel, and iron specimens that had been exposed to various benign corrosive indoor atmospheres shows large amounts of infrared bands from carboxylate ions such as formate, acetate, and oxalate. This may be an indication of a stronger influence of carboxylic acids and other organic compounds under benign conditions than in more aggressive atmospheres.

REFERENCES

1. JE Svensson, LG Johansson. A laboratory study of the effect of ozone, nitrogen dioxide, and sulfur dioxide on the atmospheric corrosion of zinc. J Electrochem Soc 140:2210, 1993.
2. WH Abbott. Corrosion of electrical contacts: review of flowing mixed gas test developments. Br Corros J 24:153, 1989.
3. American Society for Metals. Metals Handbook, Vol. 1, 8th ed. Metals Park, Ohio, 1961.
4. International Nickel Company. Properties of Some Metals and Alloys. New York, 1968.
5. National Association of Corrosion Engineeers, Laboratory Corrosion Testing of Materials for the Process Industry. NACE Standard TM-01-69, Houston, Texas, 1969.
6. J Sickfeld. Experimental aspects of adhesion testing—An IUPAC study. J Oil Color Chem Assoc 61:292–298, 1978.
7. WK Asbeck, The measurement of adhesion in absolute units by knife-cutting methods: the hesiometer. Presented at the Eleventh FATIPEC Congress, Brussels, 1972, Congress Book, pp 78–87.

8. AM Gent, GR Ramed. Peel mechanics. J Adhes 7:91–95 (1975).
9. AD Crocombe, RD Adams. Peel analysis using the finite element method. J Adhes 12:127–139, 1981.
10. PM Heertjes, J deLong, The peeling-off of paint film. J Oil Color Chem Assoc 35:1096–1106, 1972.
11. U Zoril, Testing of the impact resistance of coatings under aspects of deformation energy. Presented at the 15th FATIPEC Congress, Amsterdam, 1980, Congress Book, Vol. 3, pp 94–109.
12. PA Engel, DD Roshon. Indentation-debonding of an adhered surface layer. J Adhes 10:237–253, 1979.
13. HD Conway, JPR Thomson. The determination of bond strength of polymeric films by indentation debonding. J Adhes Sci Technol 2:227–236, 1988.
14. PA Meyer, JL Rose. Modeling concepts for studying ultrasonic wave interaction with adhesive bonds. J Adhes 8:107–120, 1976.
15. HG Mosle, B Wellenkoetter. Acoustic emission on painted steel sheets. Presented at the European Federation of Corrosion Conference on Surface Protection by Organic Coatings, Budapest, 1979, Proceedings, 5, pp 17–33.
16. W Mielke. Application of acoustic emission analysis for evaluation of the adhesion of coatings. Presented at the 19th FATIPEC Congress, Aachen, 1988, Congress Book, pp 471–486.
17. BE Dom, HE Evens, SM Torres, Thermographic detection of polymer/metal adhesion failures. In: KL Mittal, ed. Adhesion Aspects of Polymeric Coatings. New York: Plenum Press, 1983, pp 597–621.
18. J Wolstenholme, Corros Sci 13:521, 1973.
19. I.S.A.–s 71.04–1985 Standard: Environmental conditions for process measurements and control system: airborne contaminants, Instrument Society of America, Research Triangle Park, South Carolina, 1985.

Index